Whole Protein Vegetarian

Whole Protein Vegetarian

DELICIOUS PLANT-BASED RECIPES
WITH ESSENTIAL AMINO ACIDS
FOR HEALTH AND WELL-BEING

Rebecca Miller Ffrench

Photographs by Joshua Holz

The Countryman Press
A division of W. W. Norton & Company
Independent Publishers Since 1923

For information about permission to reproduce selections from this book, write to
Permissions, The Countryman Press, 500 Fifth Avenue, New York, NY 10110

For information about special discounts for bulk purchases, please contact
W. W. Norton Special Sales at specialsales@wwnorton.com or 800-233-4830

Manufacturing by Quad/Graphics Taunton
Book design by Jeri Heiden, SMOG Design, Inc.
Food styling by Rebecca Miller Ffrench and Rebecca Shim

The Countryman Press
www.countrymanpress.com

A division of W. W. Norton & Company, Inc.
500 Fifth Avenue, New York, NY 10110
www.wwnorton.com

Library of Congress Cataloging-in-Publication Data

Names: Ffrench, Rebecca Miller, author.
Title: Whole protein vegetarian : delicious plant-based recipes with
essential amino acids for health and well-being / Rebecca Miller Ffrench.
Description: Woodstock, VT : The Countryman Press, a division of
W.W. Norton & Company, [2016] | Includes bibliographical references and index.
Identifiers: LCCN 2015046862 | ISBN 9781581573268 (hardcover : alk. paper)
Subjects: LCSH: Vegetarian cooking. | Amino acids in human nutrition. |
Vegetarianism. | LCGFT: Cookbooks.
Classification: LCC TX837 .E37 2016 | DDC 641.5/636—dc23
LC record available at http://lccn.loc.gov/2015046862

10 9 8 7 6 5 4 3 2 1

TO MY MOM AND DAD FOR ALWAYS PUTTING VEGETABLES ON MY PLATE.
TO JOSH, JEN, AND GEORGE FOR YEARS OF MENTORING AND COLLABORATION.
TO CAMILLA, ANNA, AND JIM—YOU MAKE IT WHOLE.

Contents

Introduction

The kitchen is the focus of our home. We never meant it to be because it's probably the smallest room in the house. Yet it's the place where everyone hangs out, perched on stools at the well-worn counter, careful not to back up more than a foot lest they hit the wall—but that's okay, because it's still the best seat in the house. That's the place where you'll most likely be fed.

We do have a dining table where we eat many meals, but when someone in our house is cooking, the others always gather around that familiar counter. Maybe to prep some onions, or perhaps to help with dishes, but always to get a taste of what's cooking. And it's not always me at the helm. My husband, Jim, is an excellent cook, and when my girls, Anna and Camilla, take the reins, I'm buffaloed by their kitchen prowess.

Hours spent in this small room have become our family time—eating, talking, laughing, and learning. It is here, and around the dining table, that we share ideas and thoughts. It all happens around food.

Food is about community, and that's why I cook. To feed myself, but more so, to nourish my family and friends. We don't live in a mansion, and we don't live on a farm, but our house is always open to visitors. Jim and I love to entertain, and we work together to prepare big meals using a multitude of fresh ingredients. Jim inspires creativity because he pushes the envelope to explore undiscovered flavors and new methods of cooking. It's fun to see how people respond to these dishes. We thrive on the comings and goings of guests rich with knowledge and new ideas.

Over the years I've learned that sharing a meal is not about the perfect presentation nor elaborate haute cuisine—it's about sincerity. Food that is prepared simply but deliciously is genuine. And it's about love. Not just a love for the people with whom you share your food, but for the food itself. I can't even begin to tell you how my heart surges at the first sight of plump sour cherries in summer. Or how my mood is elevated by the heady scent of freshly picked mint. I have such an enormous appreciation for good, fresh ingredients.

But that wasn't always the case. Fifteen years ago when Jim and I would plan dinners, I'd want to know exactly what we were preparing *before* we went to the store. He'd say, "I'm just going to get what looks good." In essence, he was buying what was freshest, what was in season. Me, I wanted to eat what I wanted, when I wanted. Then I understood why he shopped how he did (and still does). Buying food seasonally is essential to good eating. Why in December would I eat unripe, flavorless strawberries that are white inside, when in June I can savor a deep red one

with flesh so sweet and sumptuous its juices run down my chin?

My grandmother and mother always cooked this way, using seasonal, minimally processed, natural foods. It was their love of good food that brought people to the table. Over the years, though, there was a shift away from whole foods. Not for my grandmother, because she still doesn't buy convenience foods, but for me. I always felt I ate pretty healthfully, but I really didn't. I relied a little too regularly on white pasta, dry cereals, and white-flour baked goods. Sure, the cereals were organic and the baked goods were homemade, but they were still made from refined flours and sugars.

A SHIFT IN HOW WE EAT

Then came a time when my two girls started to ask for salads. They were young, eight and eleven, but they knew. Their bodies were telling them they needed to replace refined sugars with whole foods, and to them, raw greens from the garden made them feel good.

The girls had become aware of the relationship between what they ate and how they felt. It was at that time our family really started the conversation about healthy foods.

Our kids have good appetites, and have always eaten the vegetables I served, but now they were *asking* for them. I started to focus on whole food alternatives. I substituted brown rice for white (or sometimes mixed the two), then started using other whole grains, such as barley and quinoa.

Although home-baked goods may have gotten my girls to the table, now they were ready for more. The next few years became a time of great experimentation. I explored many new ingredients, from seeds to sea vegetables, and incorporated them into our daily meals.

Throughout all of this, my first priority has been the same: delicious food. I will not sacrifice a texture just so I can use quinoa flour nor will I add flaxseed if it adversely affects the flavor. I have worked to incorporate healthful ingredients in foods so they are nutritious, but more so, tasty. If I serve a dish that's healthy but flavorless (believe me, my family kindly suffers through my failures), then it's not satisfying. There is no joy in a dish that, while good for you, doesn't taste good,

It was because of this way of eating—seasonally, whole, unprocessed—that I started developing

recipes with vegetables, grains, and legumes as costars. My family has accepted these meals wholeheartedly. At least most of the time. Sometimes recipes are met with skepticism. When I cooked up whole freekeh for the first time, my husband neighed, insinuating that it looked like something horses eat, and my kids asked, "Wait, *what* is this?" One bite of the smoky, chewy grain, though, and they were hooked.

As I've been cooking this way, my tastes have changed. I no longer crave sweet things as I used to. My blood sugar levels are more consistent, without the highs and lows I used to experience when I ate more refined flours and sugars. This story of an improved diet is not new. It's just that I'm excited to add my own ideas on how one can put plant-based foods first.

The mindset here is quality. I choose good whole grains over white refined ingredients; I buy sea-

And so my research began: How much protein do you really need? Can you get enough from a plant-based diet? Should you be supplementing with protein powders? My questions were numerous.

As you may already know, it's easy. I feel people won't even be asking the plant protein versus meat protein question in 10 years. No longer will anyone dispute the nutritional completeness of meatless meals.

Nature has made sure we are protected against a protein deficiency, as whole grains, vegetables, and legumes are excellent sources. Legumes, in particular, are rich in fiber, high in nutrients, low in saturated fat, and cholesterol-free.

Veg-centric eating is now routine for my family. One way I've kept my family interested is by offering a wide variety of foods. Kale is good for you, and you could eat it every night . . . but if that's all you ate, you'd miss a whole fortune of nutrients supplied by the bounty of plant-based foods. Within these pages, I've included that same ample variety so you'll find plant-based foods that appeal to you, provide you with balanced nutrition, and most important, taste good. I want to share with you my excitement for food because, boy, does a good, tasty bite make me happy—and hopefully will make you happy, too.

sonal produce over vegetables waiting to ripen on a ship. But this is *not* a diet book, and I'm not here to tell you what to eat. I'm here to show you how *good* good food can be.

WHOLE NATURAL FOODS ARE ABUNDANT IN PROTEIN

Vegetables have always been a part of my repertoire, but now they take precedence on the plate. I'm not vegetarian, but my family and I eat far less meat now than ever before.

The recurring question, though, is invariably asked, "How do you get enough protein if you're not eating meat?"

If you cook this way—using fresh, whole ingredients—you will see how easy it is to get the protein you need. So, let's get started. Let's find out what foods make you feel good.

PROTEIN ESSENTIALS

First though, I want to dispel some protein myths: Dietary protein is *not* only gained from eating animal products, and animal-based protein is *not* better for you than plant-based protein.

After water, protein is the second most-abundant substance in our body. It is the major structural component of every cell in our body, including our brain. In fact, our cells *need* protein. Without it, our organs cannot exist. Even hormones and antibodies are made up of proteins. About one in every ten of our calories consumed, or 10 to 35 percent of our daily calorie intake, should come from protein. It encourages weight loss, builds HDL (good cholesterol), and increases our metabolic rate.

Proteins are made up of amino acids—think of these as the building blocks of protein. There are 22 different kinds (and more are being discovered continually) that organisms use. Our body links amino acids together in thousands of ways to form protein molecules. We can produce 13 of the 22 amino acids and must obtain the others, called essential amino acids, from our food.

Protein is protein, regardless of the source. Yes, there is a difference in the amino acid profiles of animal- and plant-based proteins but, again, one is not better than the other. Animal proteins may be more bio-available than plant proteins, but the latter offers benefits that animal proteins don't, such as fiber and phytochemicals. The proteins we ingest are broken down by our cells into their individual amino acids and then reassembled to build specific new proteins that our body needs. This process is called protein biosynthesis.

Proteins, and more so amino acids, are not only building blocks. They are extremely important to healing. I learned this firsthand when I was hospitalized earlier this year. I had a terrible burn on my leg and foot and was admitted to New York Presbyterian's Burn Unit for a week. During this time, the nurses repeated over and over, "Drink your shakes, get your protein." In addition to the three meals, the hospital provided ready-made nutrition shakes that each contained 13 grams of protein. They wanted me to drink three to four a day for increased healing.

However, when I read the ingredients, I realized I wasn't only getting protein but also all these other things: water, corn maltodextrin, sugar, milk protein concentrate, canola oil, cocoa (pro-

cessed with alkali), corn oil, soy protein isolate, and less than 0.5 percent of potassium citrate, magnesium phosphate, soy lecithin, sodium citrate, natural and artificial flavor, potassium chloride . . . and that was only the first quarter of the list.

What is all that stuff? After reading the label, I could barely stomach the shakes. The ingredients all seemed to have been extracted or created. Ugh. This made me even more focused on finding healthy, natural sources of protein. I wanted whole foods, not processed shakes.

Essentially all vegetables, grains, beans, nuts, and seeds contain some protein. The difference between animal- and plant-based proteins is usually the amount of amino acids they contain. So, while animal products provide complete proteins, most plant-based sources have some protein but are usually low in certain amino acids. Grains, for example, are low in lysine. Legumes, on the other hand, are high in lysine but low in methionine. Many cultures pair grains and legumes in traditional dishes—rice or corn and beans in Central and South America and soybeans and rice in Asia. The amino acids of the foods complement each other to create whole proteins.

The idea that certain foods are low in some amino acids was discovered in the early 1900s, but it was in the 1970s that the respected, well-known anti-hunger activist Frances Moore Lappé, author of *Diet for a Small Planet* (Ballantine Books, 20th Anniversary Edition, 1991), suggested that combining foods with specific complementary amino acids at every meal was important. However, Lappé amended the statement a decade later, reporting that it was unnecessary to tediously count amino acids like calories, combining and recording them at meals. This is good news: You can get the amino acids you need from eating different types of plant-based foods and it doesn't have to be accomplished in the same meal.

> Complementary amino acids do not need to be consumed at the same time, but may be eaten over the course of a day. So, if you eat beans at lunch and rice at dinner, it's the same for your body as eating a big bowl of rice and beans together.

Not to get all science-y on you, but protein is truly fascinating. If this amino acid breakdown interests you, read on. Otherwise, just skip forward to the recipes!

The following page contains USDA daily recommended amounts of amino acids. By comparing tofu and brown rice, you see that 2½ cups of tofu meet those recommendations as do 16 cups of cooked brown rice. This chart shows all the necessary amino acids you're getting that build proteins from two examples of plant-based foods. Of course, this is not how you should get your protein, as other nutrition would be forfeited, but I think this visual reference makes it easy to understand that plant-based foods are full of proteins.

The Center for Disease Control's average requirement of protein for women aged 19 to 70 is 46 grams per day, and 56 grams per day for men over 19 years of age.

You can use this formula to calculate the daily amount of protein the United States Department of Agriculture (USDA) recommends you need:

Your weight in kilos x 0.8 grams = number of protein grams per day (or about 0.36 grams of protein per pound)

Some sources say you may need as much as 1 gram protein multiplied by your weight in kilograms if you are getting only vegetarian protein to account for decreased protein bioavailability. Children, pregnant women, the elderly, and even endurance athletes should consult the USDA website (usda.gov) for specific recommendations.

Food	His	Try	Thr	Iso	Leu	Lys	Met+Cys	Phe+Tyr	Val
USDA recommended amounts of amino acids for average person	826 mg	295 mg	1,180 mg	1,121 mg	2,478 mg	2,660 mg	1,121 mg	1,947 mg	1,416 mg
2½ cups tofu	1,455 mg	780 mg	2,045 mg	2,480 mg	3,805 mg	3,300 mg	1,335 mg	4,110 mg	2,530 mg
16 cups cooked brown rice	1,840 mg	936 mg	2,656 mg	3,056 mg	5,952 mg	2,752 mg	2,490 mg	6,432 mg	4,240 mg

Source: Dietary Reference Intakes For Energy, Carbohydrate, Fiber, Fat, Fatty Acids, Cholesterol, Protein, and Amino Acids, Institute of Medicine of the National Academies, 2002 and 2005, THE NATIONAL ACADEMIES PRESS 500 Fifth Street, N.W. Washington, DC 20001.

Amounts of amino acids are in milligrams. His=histidine Try=tryptophan, Thr=threonine, Iso=isoleucine, Leu=leucine, Lys=lysine, Met+Cys=methionine+cysteine, Phe+Tyr=phenylalanine+tyrosine, Val=valine

DAILY MEAL PAIRINGS

....

While it's easy for me to say that protein recommendations can be met through a vegetarian diet, let me show you with numbers. This sample meal plan is not designed for cooking efficiency or frugal grocery lists; it is strictly here to show you how eating a balanced plant-based diet achieves, if not exceeds, sufficient protein intake.

The daily protein recommendations below are based on the Center for Disease Control's average requirement of protein for people aged 19 to 70, which is 46 to 56 grams per day.

However, some sources* suggest a higher plant-based protein intake for vegetarians is needed to compensate for plant protein bioavailability. This recommendation is 1g protein x your weight in kilograms.

You can see that by eating a variety of dishes, including vegetables, whole grains, nuts, seeds, and some dairy, whether you go by the recommended 46g or 59g of protein per day, these intakes are easily met.

Keep in mind, this chart does not include other foods you may serve with a meal, such as a side of rice with a curry or a topping of fruit on oatmeal, which would only increase nutrition and protein values.

....

*Protein dietary reference intakes may be inadequate for vegetarians if low amounts of animal protein are consumed
Kniskern, Megan A. et al. *Nutrition*, **Volume 27, Issue 6, 727–730**

FALL/ WINTER WEEK	DAY 1	DAY 2	DAY 3	DAY 4	DAY 5	DAY 6	DAY 7
BREAKFAST	Sweet Potato Almond Milk Smoothie 7g	Overnight Steel-Cut Oats 12g	Maple Granola Clusters with Yogurt 10g	Classic Egg and Cheese Sandwich 25g	Warming Breakfast Broth 8g	Quinoa Oat Breakfast Cookies 8g	Root Vegetable Hash with Fried Eggs 13g
LUNCH	TLT Sandwich 31g	Kale, Roasted Beet, and Edamame Salad 12g	Creamy Cashew Soba Noodles 22g	Brown Rice Balls 15g	Mushroom Frittata 11g	Favorite Tofu Bánh Mì 17g	Brown Rice and Avocado Collard Wrap 16g
DINNER	Stuffed Poblano Peppers 10g	Rich Lentil Stew with Fluffy Millet 22g	Comforting Veggie Mug Pies 15g	Sea Vegetable Brown Rice Bowl 7g	Spicy Three-Bean Chili and Cornbread 28g	Red Kidney Bean Stew 7g and Saag Paneer 15g	Roasted Acorn Squash with Quinoa 20g
SNACKS	Nut + Seed Protein Bar 5g (6 ounce serving) Greek Yogurt 14g	Red Lentil Hummus Seeded Crackers 9g Sriracha Deviled Egg 12g	Roasted Edamame and Kale 7g 1 tablespoon Peanut Butter + Celery 5g	Nut + Seed Protein Bar 5g Roasted Chickpeas 7g	14 almonds 4g ¾ cup Kefir 8g	Carrot Mash Tartine 9g (6 ounce serving) Greek Yogurt 14g	Spinach and Chickpea Spoon Fritters 7g 1 tablespoon Peanut Butter + Celery 5g
TOTAL PROTEIN GRAMS FOR THE DAY	67g	67g	59g	59g	59g	70g	61g

SPRING/ SUMMER WEEK	DAY 1	DAY 2	DAY 3	DAY 4	DAY 5	DAY 6	DAY 7
BREAKFAST	Green Tea Pea Smoothie 5g	Kale and Carrot Tofu Scramble 22g	Creamy Amaranth Banana Porridge 15g	Tarragon Egg Salad with No-Knead Dark Rye Bread 22g	Savory Cheddar Pinto Bean Muffins 9g	Carrot Millet Griddlecakes 12g	Lentil, Spinach, and Tomato Frittata 20g
LUNCH	Roasted Eggplant and Zucchini Wrap 23g	Creamy Cashew Soba Noodles 22g	Vegetable Noodles with Hempseed Basil Pesto 24g	Farro, Beet, and Pea Shoot Salad 19g	Grilled Vegetable and Fresh Ricotta Sandwich 13g	Savory Spring Crostata 17g	Vegan Zucchini Roll Ups 12g
DINNER	Summer Tomatoes with Millet and Pesto Cream 19g	Sea Vegetable and Brown Rice Bowl 7g	Crispy 3-Grain Cake with Mozzarella and Tomatoes 12g	Crispy Coconut Cauliflower Curry 11g	Beet and Cranberry Bean Farrotto 27g	Portobello Mushrooms with Freekah and Artichokes 20g	Soba Noodles in Broth with Bok Choy 10g
SNACKS	Red Lentil Hummus Seeded Crackers 9g 14 Almonds 4g	Roasted Chickpeas 6g ¾ cup Kefir 8g	1 tablespoon Peanut Butter + Celery 5g Tamari Sunflower Seeds 8g	Nut + Seed Protein Bar 5g Roasted Edamame and Kale 7g	6 ounce serving Greek Yogurt 14g	Tamari Sunflower Seeds 8g Sriracha Deviled Egg 12 g	Crunchy Sesame Seed Snacks 7g Peanut Butter Power Bites 15g protein
TOTAL PROTEIN GRAMS FOR THE DAY	60g	65g	64g	64g	63g	68g	64g

Getting Started

"May I please have some more of that?" my nephew David asked of the root vegetable hash at breakfast one morning. "I just love it," he said. My heart sang. Not because he liked my cooking, but because this kid was eating vegetables, and really good ones, such as sweet potatoes and turnips. You see, David once only ate white foods: white bread, white pasta, white cheese. Yes, turnip is snowy white, but it is also a good source of fiber and vitamin C. Tossed with some fresh herbs and paired with a buttery fried egg, this hash is simple to prepare. Roasting the vegetables sweetens them and a quick fry adds a gratifying crispy crust. This particular occasion gave me hope for anyone who claims not to like vegetables. If the vegetables are fresh and prepared to enhance their natural flavors instead of masking them, even the toughest customer can usually be sold on their divineness.

Sourcing ingredients

I admit, I once treated vegetables like second-class citizens. I would decide what meat I was going to prepare, then the vegetables would follow. Now, vegetables are my starting point. I choose what looks good when I get to the market and the vegetables I pick become the anchor for my meal.

Buying the freshest ingredients is vital to good eating. If you can buy them at a farmers' market, fantastic. There has been over a 350 percent increase in farmers' markets nationwide since 1994. The USDA estimated 8,268 markets were operating in 2014.

Understanding what you're putting on your plate is a step toward a sustainable, local food system. These buzzwords—*farm to table*, *local*, *sustainable*—are used more often than they are understood. Other terms may be misused as well: One company actually labels its salt GMO-free—hasn't anyone there noticed that salt is not a living organism? This is why it's extremely important to know exactly how these words pertain to your food purchases. At a farmers' market, you can ask the vendor how and where the foodstuff was grown and harvested. Some farmers don't want to pay for an organic certification but they still use organic practices. You can find out whether the farmer sprays, and whether those sprays are harmful. Vendors are oftentimes brimming with a plethora of knowledge.

Since foods sold at markets come directly from farmers, they are fresher and retain more nutritional benefits than their mass-market counterparts. The produce farmers deliver is not sitting in storage waiting to ripen. Because there is less distance between you and your food, fewer natural resources are used to get the produce to you. A smaller carbon footprint is left. The food is usually less processed and not waxed. You don't have to scan for unwanted ingredients, such as high-fructose corn syrup—not to mention there is no packaging label claiming health benefits. There isn't even a package, for that matter.

That said, don't fret if you can't make it to a market. I certainly always don't. Our nearby grocery store is starting to carry more and more local produce, which tells me there is a demand for fresh, whole foods.

While I'm a fervent advocate of farmers' markets, the most important thing to keep in mind here is to buy whole, unprocessed foods. Even plain frozen vegetables are better than a frozen casserole-type dish loaded with preservatives. The unprocessed vegetables from grocery and big-box stores are still that—good, unprocessed vegetables!

The recipes in this book cover a wide range of ingredients, from heirloom beans and farro to pea shoots and burrata. Never heard of these things or know you can't get them at the market near you? No worries. I offer substitutions or suggested alternatives for ingredients that are just becoming available on a national scale.

If you haven't yet but are interested in exploring new (ancient) grain varieties, seek out a health food store in your area. Most likely they sell these pantry staples in bulk. Alternatively, you can order them from an online resource. As for something like pea shoots, they're only available for a short period, usually from a farmers' market. In the particular recipe calling for them, I suggest arugula as an alternative.

Don't hesitate to take advantage of your local environment. I don't even mention ramps in the book because as far as I know they can only be foraged. I am so fortunate to be able to get them in the woods near my house—one of the benefits of living in the Catskill Mountains. I do oftentimes use them in place of scallions when they are in season, and I've included a photo (pages 6–7) because they are just so darn gorgeous. What do you get in your area that I can't get in New York? Truffles in North Carolina or Oregon perhaps? I believe there are lots of hazelnuts in Oregon, too. And how about pawpaw in Ohio? This is an example of how to use what you can get your hands on. Maybe you don't have ramps, but you can grow onions in a window box, so don't hesitate to use them.

Our house is in the country, but our garden is modest and so is our harvest. Our friends and neighbors are so kind to share their yards and yields. They call to say we can pick blueberries anytime. My girls come home with purple-stained fingers—and lips—and baskets brimming with plump, juicy berries. When the peaches are ripe, our neighbors send bushelfuls our way.

How to use this book

Now that you've got the lowdown on protein and know how much you need, let's get to the good part: eating it!

The recipes in this book are ones I feed to, and have been vetted by, my family, friends, and neighbors. The dishes I used in the photos are my own, hence a small chip or imperfection, and everything was shot right in my dining room and kitchen. I tried to photograph things as they would be when I put them on the table. I find beauty in the rustic look of a stew served from a Dutch oven or vegetables piled high on a platter straight from the roasting pan. While we were photographing this book, neighbors sent over ingredients they had in abundance for the shoot—collards, kale, apples, squash, and eggs came from Holz Farm up the street. The Reisses sent over tomatoes, garlic, and lentils. We're always borrowing something or another from each other.

This is where headnotes and cook's notes come in. These tidbits of information are what I would say to you if I were cooking with you in the kitchen, just like I'd say to my friend or neighbor, "Hey there, I love this dish because . . ." or "What about using this ingredient instead?" I might also say, "Did you know this is good for you because . . ."

My hope is that after you've followed some of the recipes in this book, you start to experiment. The idea is for you to find what combinations you like, and then expand on them using the ingredients you have on hand or that are available at the market. You can and should most definitely follow the recipes as written, but you can also use them as guidelines. Insert your own flair and favorite flavors. I've called on my travel and culinary experiences to bring varied spices and ingredient combinations to the pages of this book. From Jamaican Rice and Peas (page 165) to Soba Noodles in Broth with Bok Choy (page 131), these recipes are reflective of travels spanning from Jamaica to Japan.

When you're cooking, think in terms of your five tastes: sweet, salty, sour, bitter, and umami. If something seems flat, add some vinegar or freshly squeezed lemon juice. If it needs a fresh accent, throw in a handful of cilantro or parsley. Never shy away from salt and pepper.

Speaking of which, I use salt and pepper quite liberally. If you are on a restricted diet, of course amend the amounts to your needs. I talk more specifically about types and brands in the "Stocking Your Pantry and Fridge" section (page 17). In general though, if I say a pinch or two, it's just under 1/8 teaspoon. I use kosher salt, so a pinch of that will yield more than a pinch of table salt. For other measurements, when I call for a heaping cup, it is a cup plus a small handful.

In recipes that call for milk, I either specify whole milk for cow's milk or list the specific type of milk, such as almond milk, coconut milk, and so on. If you can use any type of milk—cow, nut, soy—in a recipe I say "milk of your choice." Keep in mind that the protein changes significantly when you use nut milks in place of cow or soy milk. Nut milks are very low in protein.

I've also analyzed the protein content of a serving of each recipe, so when you put together a day's meal plan, you'll have a general idea of how many protein grams you're taking in.

A good point of reference: a 3-ounce can of chunk light tuna in water has 16 grams of protein. Many of the lunch recipes contain around that on average.

In the dinner chapter, there are some recipes that contain less, but that is because I've written those to be eaten with other dishes. Pair several recipes together for a complete meal.

When planning the chapters, I've thought of the lunch dishes as more stand-alone, sandwich, and salad-type meals, while quite a few of the dinner recipes are meant to be paired with another dish, if even a salad. This is to provide variety of tastes and textures.

Of course, I hope you'll combine any and all of these recipes without feeling you need to follow a prescribed mealtime or plan. Have breakfast for dinner. We often do.

Also note that serving sizes are general. Most recipes serve four, but some serve more, some less. For example, I generally make smoothies for one or two. You can easily double those recipes, though. I base the recipes on what I feel would feed the average woman (like myself), so adjust portions accordingly.

Stocking your pantry and fridge

Nearly all of the ingredients in this book should be available in most parts of the country. The food landscape has changed—ingredients that I could have never sourced locally (or even known about) 10 years ago are now sold at my nearby grocery store. I try to gauge availability by what's at our small market upstate where I can now find millet, chia seeds, and farro.

Adding new healthy foods to your pantry doesn't come cheaply, though. I do find a little goes a long way. Just a splash of a complex balsamic vinegar can brighten a dish and a good jar of honey can last for months.

GRAINS

barley, brown rice, old-fashioned rolled oats, farro, freekeh, amaranth, whole buckwheat groats, quinoa

Keeping an assortment of grains in my pantry encourages me to break routine and cook with variety. While amaranth, buckwheat, and quinoa are actually seeds and not grains, I've included them here because they're treated like grains. Grains vary in taste, but can often be substituted for one of a similar size and texture in many dishes, so don't hesitate to experiment.

If your local market doesn't carry a particular grain, check out a health food store or look for a retailer that sells Bob's Red Mill products. The company offers a large selection of organic whole grains and legumes nationwide.

If gluten is an issue when you're cooking, for yourself or others, be sure to avoid barley, rye, and wheat, including all varieties such as freekeh and farro (emmer, einkorn, and spelt). Buckwheat is safe though. Also, although oats do not inherently contain wheat, they can be cross-contaminated, so use only certified gluten-free oats and oat flour.

FLOURS

white whole-wheat, spelt, oat, rye, light buckwheat, almond, quinoa, and coconut flours

If you have a high-powered blender, I encourage you to take a stab at grinding your own flours (page 37). It takes seconds, and making nut and grain flours is much more cost-effective than buying them. I am a big fan of King Arthur Flour products for both ground flours and whole grains.

I have replaced the use of all-purpose white flour almost exclusively with white whole-wheat flour in my kitchen. White whole-wheat is the same as whole-wheat flour in the sense that it's ground from the entire wheat kernel—unlike all-purpose white flour, which is bleached and the bran and germ removed. The difference is that white whole-wheat flour is ground from a hard white wheat berry instead of a red wheat berry, like traditional whole-wheat flour. The result: a flour that has all the nutritional advantages of traditional whole-wheat but with a lighter color and milder taste.

Oat and almond flours are two of my favorites for gluten-free baking.

High-Protein Whole Food Replacements for Refined Foods

Refined/Processed Food	Protein Content	Fiber Content	High-Protein Whole Food	Protein Content	Fiber Content
1 cup/125 grams enriched all-purpose flour	13 grams	1 gram	1 cup/112 grams almond flour	24 grams	12 grams
1 cup/28 grams crispy rice cereal	2 grams	0 gram	1 cup/246 grams cooked amaranth	9 grams	5 grams
1.5-ounce bag mini pretzels	4 grams	1 gram	2 tablespoons/1 ounce sunflower seeds	7 grams	4 grams
1/3 cup/62 grams white, long-grain enriched rice	5 grams	1 gram	1/3 cup/60 grams dry black rice	6 grams	3 grams
1 serving Reese's Peanut Butter Cups	5 grams	1 gram	1 serving Peanut Butter Power Bites	15 grams	5 grams

LEGUMES

chickpeas, lentils, black beans, pinto beans, adzuki beans, cannellini beans, red beans, heirloom bean varieties, edamame

Both dried and canned beans are easy to store, so I keep an abundance of them in my pantry. There is no question that dried beans are tastier than canned. The texture is firmer and the taste is more prominent. Canned beans work well though when you're pressed for time. While some say beans store indefinitely, they are best used within a year of their expiration. The longer you keep beans, the longer they take to cook (old beans could take as long as two hours).

NUTS AND SEEDS

almonds, pecans, hazelnuts, walnuts, peanuts, flaxseeds, hemp seeds, pumpkin seeds, chia seeds, sesame seeds, nut butters

Nuts and seeds are compact protein packs. These nutritional powerhouses are high in healthy fats, too, such as omega-3s. I tend to store the bulk of my nuts and seeds in the freezer though, because they can go rancid quickly due to their high fat content. I keep smaller amounts in airtight lidded glass jars. A handful of nuts or seeds sprinkled on a dish adds interest to any meal. You'll notice I include peanuts in this grouping, which I do because, although they are legumes, they are treated as nuts in the US.

DRIED MUSHROOMS AND SEA VEGETABLES

porcini, shiitake, kombu, nori, arame, hijiki

Both mushrooms and sea vegetables (seaweed) are thought to have great healing properties. Sea vegetables are good sources of minerals—iron, calcium, magnesium—and mushrooms are loaded with antioxidants. I especially like to use both of these ingredients for making broths.

OILS AND FATS

cold-pressed extra-virgin olive oil, unrefined extra-virgin coconut oil, sesame oil, a neutral oil such as canola or grapeseed, hot chili oil, grass-fed butter

Fats were once shunned and low- or nonfat diets were touted as healthy. We now know that healthy fats are essential for proper body function—fats help your body absorb fat-soluble vitamins and some high-quality oils can actually increase good HDL cholesterol in the blood.

Good-quality, cold-pressed extra-virgin olive oil is my go-to fat. I buy it by the liter and use it on most everything. I even use it for popping corn.

12 PROTEIN-BOOSTING PLANT-BASED FOODS

1 cup cooked lentils = 18 grams protein

1 cup shelled edamame = 16 grams protein

1 cup cooked black beans = 15 grams protein

1 cup cooked quinoa = 8 grams protein

3 ounces soy tempeh = 16 grams protein

3 ounces tofu = 8 grams protein

2 slices whole grain bread = 7 to 8 grams protein

2 tablespoons hemp seed = 8 grams protein

15 almonds = 4 grams protein

1 cup sun-dried tomatoes = 8 grams protein

1 cup raw spinach = 5 grams protein

1 whole avocado = 4 grams protein

I once made popcorn with canola oil and my kids asked why it wasn't good. Aah, no olive oil. Oils add flavor and rich texture to foods when you use them. We are satiated when we get the fats we need.

Substituting a few protein- and fiber-rich foods for less nutritious ingredients commonly found in pantries can make a big difference.

Coconut oil is another favorite, as is organic butter, made from the milk of grass-fed cows. There really is no replacement for the taste of creamy butter. But for all fats, moderation is key.

SALTS, HERBS, AND SPICES

salts include Diamond Crystal kosher salt, Maldon sea salt; herbs include fresh thyme, tarragon, rosemary, basil, cilantro, flat-leaf parsley, oregano, and dried bay leaves; dried spices include cumin, coriander, cinnamon, turmeric, paprika, garam masala, chili powder, cayenne pepper, and black peppercorns in a grinder

Proper seasoning heightens the flavor of foods. I used to be hesitant with salt—now I find it enlivens flavors, as does the liberal use of fresh herbs. If you find a dish is missing something, often it can just be the lack of salt. I am a Diamond Crystal kosher salt devotee. It has a wonderfully coarse texture that allows you to control the amount when adding a pinch, which is hard to do with table salt. It also has no additives.

I always try to have some fresh herbs on hand. Just a sprinkle can elevate most any vegetable a notch. If I could only grow one thing in my garden, herbs would be it. Don't ask me to choose which one though . . . I don't know that I could. Basil? Thyme? Rosemary? I buy spices I don't use often in small amounts to ensure freshness and aromatic flavors.

I'd be remiss if I didn't mention the importance of a pepper mill. There is no replacement for freshly ground pepper. Pepper that is purchased already ground seems absolutely tasteless. A grinder that you can use to crack the peppercorns right over your food is key. I don't think there's a day that goes by that I don't use my pepper mill (unless I eat all my meals out!).

VINEGARS

balsamic vinegar, cider vinegar, umeboshi vinegar, rice vinegar

It can be tempting to buy the cheapest vinegar on the shelf, but with this pantry staple, you most definitely get what you pay for. For example, true balsamic vinegar is concentrated, fermented, and aged. The result is a dark, rich, and thick vinegar to be used sparingly. Likewise, raw, unfiltered, unpasteurized vinegars are very different from the refined and distilled ones usually found in most supermarkets. When vinegar is pasteurized, it is heated to a degree that beneficial nutrients are removed. While those refined vinegars are fine for cleaning, look for Bragg brand cider vinegar or other naturally fermented ones to use when cooking.

SWEETENERS

pure maple syrup, honey, coconut sugar

Sugar is sugar is sugar when it comes to calories. Whether it's honey, brown sugar, or molasses, they're all quite comparable. There's so much controversy regarding sugar these days. My thought: Use less. When I do use sweeteners, I prefer maple syrup, honey, or coconut sugar because they are close to their original source. I use very little granulated sugar anymore.

DAIRY AND EGGS

milk, buttermilk, cream, yogurt, butter/ ghee, ricotta, eggs

When it comes to dairy products, I try to buy closest to the source whenever possible. I now seek out cheeses at the farmers' market and always try to buy natural dairy products that are made from grass-fed, pasture-raised cows.

Organic products can be outrageously expensive though, so I use them in moderation (but still splurge on higher quality for more nutritional value—and taste!). Compare, for example, butter from grass-fed cows versus grain-fed ones. The butter from cows who eat grass has more vitamins and omega-3s, plus it's golden, creamy, and delicious.

Likewise, I seek out eggs from pastured chickens, ones that are out running in fields. Again, a farmers' market is a good source for fresh eggs.

DAIRY AND EGGS PROVIDE HIGH-QUALITY PROTEINS

1 cup Greek-style yogurt = 20 grams protein
1 cup plain yogurt = 12 grams protein
1 cup kefir = 11 grams protein
1 cup whole milk = 8 grams protein
1 ounce Parmesan cheese = 10 grams protein
1 ounce Cheddar cheese = 6 grams protein
¼ cup ricotta cheese = 7 grams protein
1 large raw egg = 6 grams protein

KITCHEN TOOLS TO CONSIDER

As for kitchen tools, I have listed the 12 items I use daily. Of course, I could make do with less, but these are what I consider my essentials.

No fancy equipment or tools are required for recipes in this book. However, I do use a trendy little spiralizer for one recipe, and if you have a mandoline, that could come in handy, too.

The three appliances I use again and again are a food processor, blender, and stand mixer. In addition to these items, I'd say good knives and cutting boards are two of the best investments I've made.

CAST-IRON SKILLET

If I could take one kitchen tool to a desert island, this would be it—inexpensive yet sturdy. Most anything can be cooked in a cast-iron skillet, from stir-fries and sautés to breads and pancakes. I kind of wonder how I ever lived without one. When treated properly, a seasoned cast-iron pan has a wonderful nonstick surface that is great for searing. I never use soap when washing it. I just scrub it clean with hot water and heat it on the stovetop immediately until it's bone dry, to prevent rust from developing. Never let your cast-iron sit in water. If you can't clean it right away, you're better off leaving it to sit with food debris until you're ready to clean it.

DUTCH OVEN

A good-quality enameled or cast-iron Dutch oven is a beautiful thing. It can go from stove to table; it conducts heat well; you can fry in it, braise in it, and make soups in it.

MY 12 MOST-USED KITCHEN TOOLS

Cast-iron skillet
Dutch oven
Two sharp knives
Wooden cutting board
Microplane grater
Colander/strainer
Glass bowls
Wooden spoons
Citrus juicer
Kitchen scale
Good baking sheet
Parchment paper

TWO SHARP KNIVES

Good knives are critical to success in the kitchen. Have you ever tried cutting a squash with a dull knife and had that knife slip? It's really scary. A dull knife is actually a hazard just for that reason. Professional chefs carry their own knives so they can keep them sharp and cared for. Keeping a knife clean and dry is extremely important to the life of the blade. Knives should never be stored loosely in a drawer. Either a knife guard or sheath works if you don't have a knife block or rack for storage.

While a cast-iron skillet is my true kitchen darling, I'm very fond of our Sabatier chef's knife. This high-quality knife is well worth the expensive price tag. We have accumulated several knives over the years, including a boning and serrated knife, but my second most used knife is a paring knife, good for peeling and sectioning fruits.

A knife sharpener is essential too. It need not be fancy. A small and inexpensive handheld one does the job just fine.

WOODEN CUTTING BOARD

I prefer wooden cutting boards over plastic because I feel they're gentler on knives and respond better to a blade. I have a number of them and use one specifically for onions and garlic. I've made the mistake of cutting a sweet tart on a board that was used for onions earlier in the evening, which although it had been washed, still made dessert taste oniony.

MICROPLANE GRATER

There are now many different types and sizes of Microplane graters. I have several, but the one I use most is the classic thin zester/grater. It works for lemon zest as well as finely grated Parmesan.

COLANDER/STRAINER

I like a good, sturdy colander with a base and handles for rinsing vegetables and draining beans and pastas, but I use my long-handled fine-mesh strainer for most everything else: straining nut milks, sifting flours, draining grains, and making cheese.

GLASS BOWLS

A large set of glass bowls was not a purchase I expected to appreciate as much as I do. They work for many different purposes with prepping, mixing, or combining a large number of ingredients. They're also compact because they nest, so they fit nicely in a small cupboard space.

WOODEN SPOONS

There is a large crock next to my stove jammed with spatulas, whisks, and wooden spoons of many sizes that I've collected over the years. I once stayed in a rental house without wooden spoons—only plastic ones—and was surprised at how much I missed their sturdier cousin. A wooden spoon is strong and feels good in the hand, plus, although they can burn if put near flames, they don't melt.

CITRUS JUICER

I use a simple wooden reamer, which is basically a handle with a pointy, tapered, ribbed spherical cone attached. I have three other kinds of juicers and never bother to use them as this little hand-held tool is the quickest and most efficient.

KITCHEN SCALE

Cooking by weight is important when preparing food in large quantities, especially when baking. For this book, though, I use cup and spoon measures because it is the standard for most US kitchens. I do try to include weights of foods that are not measured by cups; for example, whole eggplants or potatoes. I have a digital scale that cost less than $30 and serves me very well.

BAKING SHEET

There is no replacement for a good baking sheet. Cheaply made pans may not heat evenly and may cause foods to burn. Thick aluminum, rimmed pans (11 x 17 inches) are my absolute favorite. In fact, I have no others. I roast directly on the pans and they have developed a nice patina, which gives them a nonsticklike finish. I have two pans that I keep just for baking, almost always with parchment or a silicone mat, so they remain pristine.

PARCHMENT PAPER

Baking sheets and parchment paper go hand in hand. Lining trays and pans with parchment prevents foods from sticking to them when cooking and can also aid in lifting baked goods out of pans. It makes clean up really easy too. You can also use parchment as its own self-contained baking pan (see page 159, cooking *en papillote*). I always buy two rolls at a time so that I'm never without.

Cooking Techniques

While I truly believe whole, fresh ingredients are the secret to good cooking, mastering some basic techniques can make food preparation much more enjoyable. Cooking a perfect pot of beans comes strictly from practice—knowing to check them after a certain amount of time, continuing to test them until they're done, and not worrying if they still need to cook after an hour of simmering time. This kind of confidence comes from cooking beans again and again. Just like roasting vegetables. You'll recognize the familiar smell of carrots caramelizing and know it's time to toss them. No recipe needed. The following chapter provides a quick lowdown on a handful of techniques I use daily. These are not novel and certainly an experienced chef will be familiar with them, but they are essential to many of the recipes in this book. You'll find other cooking methods and preparations, from quick pickling to frying, described within the recipes, but the following how-tos will give you a good start.

GRAINS

Whole grains, as opposed to processed grains, are the most beneficial to any diet. They still contain the bran and germ, which is where most of the protein and other nutritional content is stored. The outer layers of a grain kernel, including the bran, are removed from processed grains.

Whole grains do have a higher fat content than processed grains, which means the oils can go rancid faster (which is one reason processed grains gained such popularity—their long shelf life).

COOKING TIMES

Grains can be soaked to shorten cooking times, but overall, most grains are cooked on the stovetop with a 2:1 liquid-to-grain ratio. More liquid may be needed for particular grains, such as barley. Generally, grains and liquids are brought to a boil, then the heat is lowered to a simmer, the grains are covered, and they are cooked at a low temperature until tender.

Cooking times depend on whether the grain is whole, hulled, pearled, polished, or rolled, among other processes. For actual cooking times, refer to the package or the directions available in the bulk grain section for the particular grain you are preparing.

Sturdier grains, such as rice, barley, and farro, can be cooked in big batches and frozen in dinner-size portions to have on hand for quick weeknight meals.

TO SOAK OR NOT TO SOAK

Nuts, seeds, grains, legumes—I use loads of them in my cooking, and in this book. They all naturally contain phytic acid. Phytic acid binds to minerals and prevents them from being absorbed by our body. It's debatable whether phytates (phytic acid) should be of concern. However, to avoid them, you can soak nuts, seeds, grains, and legumes. Not only does soaking reduce phytic acid, it also neutralizes enzymes, which some say can cause trouble with digestion.

DRIED BEANS, NUTS, AND SEEDS

COOKING TIMES

Unlike grains, which are pretty standard in their cooking times, dried beans can take anywhere from 45 minutes to 2 hours (or possibly even longer) to cook.

Soaking beans overnight not only reduces cooking times but it also breaks down gas-producing compounds in beans. Beans can double or triple in size when soaked, so put them in a large container and cover them with ample amounts of water. Pick over your beans for pebbles or other debris before soaking them and remove any beans that float while soaking them.

After you have soaked the beans overnight (or for 8 hours), rinse and drain them well. Put them in a cooking pot and add enough water to cover them by at least an inch or two. At this point you can add a bay leaf and even a mix of diced carrot, celery, onion, and herbs tied in a cheesecloth for flavor. Bring the water to a boil. Lower heat, cover the pot with a lid, and simmer gently until tender.

There is no exact science to cooking beans. Lentils tend to take only 30 minutes or less to cook, whereas dried kidney beans can take up to 2 hours. After about 40 minutes, check your beans periodically, about every 10 minutes, to see whether they are tender. When they are, drain off any excess liquid and use as desired. Conversely, if the beans look as if they may boil dry at any point, add some extra water or stock. Some recipes do require you to save the cooking liquid, though, so check before discarding it.

Add salt when the beans just start to become tender. If you add salt at the beginning of the cooking process, it can cause them to become tough.

SPROUTING

While sprouting may sound intimidating—because if you don't have a green thumb, growing anything can seem daunting (believe me, I know)—sprouting beans and seeds is actually a lot of fun. There are no weeds to pull, either!

You can sprout most type of beans, including adzuki beans, peas, and lentils. You can also sprout seeds, such as pumpkin. I call for sprouted mung beans in Crunchy Mung Bean Sprout Salad

(page 93), so mung beans could be a good place to start. Sprouts can take anywhere from 3 to 5 days to grow, but most grow within 2 days. The most important element when growing sprouts is to rinse them at least two times a day.

To sprout beans, put ½ cup or so of dried beans in a large bowl and cover them with several inches of water. (The yield is at least two times, if not four to five times, their original size.) Cover the bowl with a plate or towel and leave the beans to sit overnight.

In the morning, drain and rinse the beans. They are now activated. At this point you may keep the beans in a strainer over a bowl covered with a light cloth, continuing to rinse them every 6 hours or so (two times a day minimum). You may also do this same process in a jar covered with cheesecloth secured with a rubber band. Fill the jar with water and turn it upside down to drain out the water through the cheesecloth.

Over the next few days, sprouts will start to appear and could be anywhere from 1/8 inch to 2 inches. Keep the sprouts refrigerated for up to 2 days.

TOASTING

Toasting nuts and seeds heightens their flavor because, like whole grains, they have a high oil content. Toasting also makes them crunchier.

The most common ways to toast nuts are in the oven or on the stovetop. I prefer the stovetop because you can better control the end result.

To toast them on the stovetop, place the nuts or seeds in a dry skillet and heat over medium heat for several minutes (usually 3 to 4 minutes), stirring constantly or continually shaking the pan.

Because of the high fat content, nuts and seeds tend to burn easily. Stop as soon as you see the nuts or seeds turning slightly golden—or when you smell a fragrant, toasted aroma. Use your sight and nose to judge doneness.

Transfer the toasted nuts or seeds immediately to a bowl or plate, as they will continue to cook in the heated pan even if it's off the heat.

Vegetables

ROASTING, SAUTÉING, AND STEAMING

Gone are the days of boiling vegetables until they are unrecognizable. Vegetables are now the stars and cooking techniques should only heighten their flavors. I favor roasting. It is relatively easy, can be fast, and deepens their taste. The oven's heat caramelizes the vegetables' skins while it softens and sweetens their interiors.

I routinely toss sliced or chopped vegetables on an 11 x 17-inch rimmed baking sheet with a tablespoon or two of olive oil and a heavy pinch or two of kosher salt. I add a few grinds of black pepper and roast at 400°F for anywhere from 15 to 40 minutes, depending on the vegetable. I toss them occasionally with a spatula to ensure even cooking. You will find this method used religiously throughout the book.

Second in frequency to roasting is sautéing, which I do almost as often. The base of many of my savory recipes starts with sautéed onions, garlic, and carrots or other vegetables, such as leeks or pepper, a mirepoix of sorts.

Steaming vegetables, or cooking them in a steaming basket over rapidly simmering water, cooks them quickly while retaining their nutrients. I prefer this method to boiling because it preserves their flavor better, too.

With all of these techniques, I find the more I use them, the more they become second nature. If these are new to you, with practice, you will soon know the cooking times for your oven and pans and will be able to whip out some roasted veggies without looking at a book.

HOW TO ROAST PEPPERS

Roasting red peppers is something I do to

deepen their flavor and remove their skins. To do this, turn on the broiler. Next, cover a large, rimmed baking sheet with parchment paper. Lay the whole peppers on the prepared pan and brush with olive oil.

Place the pan with the peppers under the broiler and after 5 minutes or so, areas of the skin should start to char. When this happens, flip the peppers, using tongs. Continue to broil them on the other side until the skin there blisters as well.

After about 8 minutes, remove the peppers from the broiler and transfer them to a heat-proof bowl. Cover the bowl with a plate and allow the peppers to sit. After 10 minutes or so, the skins will slip right off. Cut them into quarters and discard the stems and seeds as well unless you are stuffing them, in which case keep them whole. Use immediately or keep refrigerated in an airtight container for up to 2 weeks.

HOW TO ROAST SWEET POTATOES

Not only are whole, roasted sweet potatoes delicious to eat right out of the skin, but their soft, sweet insides make an incredibly nutrient-dense addition to many dishes, such as the

Sweet Potato Almond Milk Smoothie (page 50) or Roasted Sweet Potato Pudding (page 192).

To roast potatoes, preheat the oven to 400°F. Next, line a baking sheet with foil. Wash and dry the potatoes. Using a sharp fork, pierce the potatoes all over and place them on the baking sheet. Put them in the oven to roast. Most potatoes take about an hour but some may be done sooner. After 40 minutes insert a fork into the potatoes to check for doneness. They should feel soft all the way through. If not, check in another 10 minutes and continue to do so until cooked through. Remove from the oven and use as desired.

HOW TO SAUTÉ GREENS

To sauté greens, heat 1 tablespoon olive oil in a large skillet over medium heat. Add one minced garlic clove and sauté for a few seconds, then put the washed and trimmed greens into the pan and stir until the greens wilt. Sprinkle liberally with salt and pepper. You may top with a good squeeze of lemon, sesame seeds, a dash of tamari sauce, or even a sprinkling of balsamic vinegar. Toasted nuts are always a good addition too. Serve immediately.

Making Homemade Staples

These are a few pantry and kitchen staples that I always have on hand. They are surprisingly easy to make. When you make foods like this from scratch, you're guaranteed a fresh product with no preservatives that is cheaper than its store-bought counterparts. Plus, the taste . . . oh, the taste. No jarred mayonnaise compares to creamy, garlicky aioli (at least not one that I've found yet), and smooth, spreadable homemade ricotta trumps the rubbery, mass-market cheese any day.

You'll find the following staples called for in recipes throughout the book. However, life gets in the way and there isn't always time for an extra step. Don't think for a moment that buying these foods ready-made will compromise the recipes in any way. Just know the recipes are here if you want them.

Roasted Vegetable Stock

I've used packaged vegetable stocks only to find they've ruined my dish because of an off-putting taste. For store-bought stocks, find a brand you like and stick with it. Vegetarian bouillons can work well, too. With homemade stock, though, you're always guaranteed a rewarding flavor. The roasted vegetables here give rich undertones to this kitchen staple. When making this stock, there's no need to peel the vegetables; just rinse them, including the onions and garlic, as they all get strained out.

Put all the vegetables on a rimmed baking sheet and toss with the olive oil. Roast at 400°F for 45 minutes. Transfer the vegetables and any juices to a stockpot and add the water and the kombu, rosemary, thyme, and bay leaves. Bring the water to a boil, then remove the kombu and lower the heat. Simmer, partially covered, for 45 minutes to an hour. Strain. Use immediately or store refrigerated for up to 1 week or in the freezer for up to 3 months.

Makes 4 to 6 cups stock, depending on how long you simmer it

2 onions, quartered

1 shallot, quartered

3 celery stalks, cut into 3-inch pieces

2 large carrots, cut into 3-inch pieces

4 ounces mushrooms

1 small fennel bulb, quartered

½ garlic head

1 tablespoon extra-virgin olive oil

10 cups water

1 (¼-ounce) piece kombu (4 x 6-inches long)

2 rosemary sprigs

2 thyme sprigs

2 bay leaves

Vegetarian Shiitake Dashi (Japanese Stock)

A clear sea stock using kombu (a type of dried kelp) and dried mushrooms instead of bonito flakes (dried fish), which are used in traditional dashi, this base for soups is extremely easy to make.

Makes about 6 cups stock

8 cups water at about 110°F-120°F
2 (¼-ounce) pieces kombu (about
    4 x 6-inches long)
1.5 ounces dried shiitake mushrooms

Put the kombu and mushrooms in a stockpot with the warm water and soak for 1 hour. After that time, turn on the heat and bring the water to a boil. Just when the water comes to a boil, remove the kombu with a large fork and simmer the broth for 5 minutes. Strain the broth and reserve the mushrooms for another use. Use immediately or store refrigerated for up to 1 week or in the freezer for up to 3 months.

Nut and Grain Flours

Flours are astonishingly simple to make, so instead of writing out the directions as a recipe, I'm just going to tell you how to make them. You put the nut or grain from which you want to make flour into a high-power blender jar and run it on high speed for 30 to 60 seconds. Scrape the sides down with a spatula and run it once more quickly—that's it, you're done. Easy, right? Just to name a few types of flour you can make: Use almonds to make almond flour, dried quinoa for quinoa flour, uncooked old-fashioned rolled oats for oat flour, and wheat berries for wheat flour. In general, 1 cup of whole nuts or grains yields just over 1 cup of flour, usually the amount is increased by about ¼ cup.

Nut Butters

Nut butters are the ideal nutritional spread. Just one tablespoon is packed with protein, fiber, and good fats. Fancy nut butters, such as almond and cashew, have gotten expensive in stores, though. If you have a blender or food processor, it's much more economical to buy nuts in bulk and grind them yourself.

Makes about 1 ½ cups nut butter

3 cups raw nuts, such as almonds or cashews, or peanuts

½ teaspoon kosher salt

2 tablespoons canola or coconut oil, melted, plus more for rubbing the blender

1 to 2 teaspoons honey (optional)

Preheat the oven to 350°F. Spread the nuts in a single layer on a parchment-lined baking sheet. Bake them for 10 to 15 minutes, or just until they begin to brown.

Rub the inside of a blender jar or food processor with oil and place the cooled nuts inside. Process for 20 seconds (on low speed if you're using a blender). Add the salt, oil, and honey, if desired. Then, process or blend again for about 2 more minutes, until the nuts become smooth and creamy. Add more oil if needed.

Store refrigerated in an airtight container for up to 4 months.

Nut Milks

Making nut milks is extremely easy. However, a good blender is required. An excellent alternative to dairy milks, nut milks are low in carbs, easy to digest, and cholesterol- and lactose-free. Almond milk is my personal favorite, but you can use any type of nut, including macadamia or pine nuts. When replacing dairy milk with nut milks though, keep in mind that nut milks contain very little protein.

Makes about 3 cups milk

1 cup nuts, such as almonds or cashews, soaked for 8 hours, then drained, soaking water discarded
3 cups water
Sweetener such as honey or maple syrup (optional)

Place the nuts and water in a blender jar and process on high speed until completely smooth. Add sweetener, if desired. Strain the mixture, using a nut bag or strainer. Discard the pulp. Chill and serve. Keep refrigerated and use within several days.

Ghee (Clarified Butter)

Ghee is a type of clarified butter traditionally used in Indian cooking. To clarify butter, you remove the milk solids by separating them out with heat. Because the milk and protein solids are removed from ghee, it may be easier to tolerate than butter for those sensitive to dairy products. While some stores sell ghee now, it is very easy to make at home. I love cooking with this glorious golden liquid fat, which has a high smoking point.

Put the butter in a small saucepan and heat over low heat until the butter melts and the solids start to separate, 5 to 8 minutes. Continue to cook slowly until foam appears. You may need to turn up the heat slightly to medium-low for this to happen. After 8 to 10 minutes, the solids will turn brown and start to settle on the bottom of the pan. At this time, remove the pan from the heat, skim off any foam, and strain the butter through a cheesecloth or fine-mesh strainer. You will yield a clear, golden liquid. Discard the solids. Store in a lidded jar at room temperature or in the refrigerator for up to 1 month.

Makes ¾ cup ghee

½ pound good-quality unsalted butter

Garlic-Infused Oil

For people with a garlic sensitivity, or those following a low-FODMAP diet, garlic-infused oil is a good option if you want the flavor of garlic without the intestinal distress eating the actual garlic can cause. This oil is great in salad dressings.

Pour ¼ cup of the olive oil into a skillet and heat it over medium heat. Add the garlic cloves and bring the oil to a gentle simmer. Add the remaining cup of olive oil and cook over medium heat for 3 to 4 minutes. Turn off the heat and let the oil cool.

Strain the oil and use within 24 hours. Or you may store it in a sterilized container in the refrigerator for up to 1 week.

Makes 1¼ cups oil

1¼ cups extra-virgin olive oil, divided
8 garlic cloves, peeled and smashed

Lemony Aioli
(Homemade Olive Oil Mayonnaise)

Slightly thinner than traditional mayonnaise, this condiment is rich yellow, with the distinct taste of olive oil. You can dress it up with fresh herbs, spice it up with sriracha, or even use all canola or grapeseed oil and no garlic for a more neutral flavor.

Makes ½ cup aioli

1 garlic clove, peeled
2 large egg yolks
½ teaspoon kosher salt
1 tablespoon fresh lemon juice
2 tablespoons water, divided
¼ cup canola or grapeseed oil
½ cup extra-virgin olive oil
1 teaspoon lemon zest

Put the garlic clove, egg yolks, salt, lemon juice, and 1 tablespoon of the water in a blender jar and process on medium speed for 5 seconds. Using a spatula, scrape the sides of the blender and push the ingredients toward the blade.

Remove the plug from the center of the blender lid and with the blender running on medium, slowly pour the canola oil drop by drop through the lid. You may want to use your hand as a shield so the mixture doesn't splatter all over. Continue adding all the canola oil this way. It should take about 2 minutes.

Add the second tablespoon of water and blend on medium speed for 5 seconds. Begin adding the olive oil drop by drop as well. Once the mixture starts to emulsify, you may pour the remaining oil in a steady stream until all of it is incorporated. Stir in the lemon zest. Refrigerate until ready for use, up to 3 days.

Tahini

Also known as sesame seed paste, tahini is similar to a nut butter in texture, and essentially made the same way. It is used often in Middle Eastern cooking. Store-bought tahini is usually made of hulled seeds, from which the hull is removed. Unhulled seeds have the hull intact and are a better source of calcium and iron. Unhulled seeds don't get quite as smooth, though, when ground. Use tahini in hummus, salad dressings, and on sandwiches.

Toast the seeds in a 12-inch skillet over medium heat for 3 to 4 minutes, shaking the pan vigorously or stirring with a spoon to prevent burning, which can happen quickly because they are so small.

Transfer the seeds immediately to the bowl of a food processor and allow them to cool (they may continue to cook if they stay in the hot pan). Once they've cooled, run the food processor for a minute or two, or until the seeds become crumbly. Add the oil and pulse 6 to 8 more times, or until the mixture is smooth and creamy, scraping the sides of the bowl with a spatula as necessary. Store refrigerated in an airtight container for 6 months or more. Bring it to room temperature before using, so you can stir in any oil that has separated out.

Makes ½ cup tahini

1 cup sesame seeds
3 tablespoons sesame oil

Whole-Milk Ricotta Cheese

After I had my first taste of fresh, creamy ricotta, I swore off the mass-market rubbery stuff forever. Hands down there is absolutely no comparison between the two. It may be difficult to find freshly made ricotta in stores, though. Luckily, this is really, truly one of the easiest staples to make—you simply separate the curds from the whey—so we can all enjoy homemade ricotta.

I rave about the ricotta, or curds, but what about the whey? It seems criminal to discard the liquid that is drained off. Homemade whey contains essential amino acids that are highly available to form complete proteins. Use this nutrient-rich liquid in smoothies, to soak beans or grains, thicken sauces, or even for baking.

Makes about 1 cup ricotta

8 cups (1 half-gallon) whole milk
¼ cup heavy cream
¾ teaspoon kosher salt
¼ cup fresh lemon juice

Line a fine-mesh colander with a double layer of cheesecloth and set over a bowl. In a large pot, bring the milk, cream, and salt barely to a boil over medium heat, continuously stirring so it doesn't burn. When the milk is just about at a boil, add the lemon juice and immediately reduce the heat to low. Continue to stir until curds form, about 1 more minute. Remove the mixture from the heat and cover. Allow it to stand for 10 minutes for the curds to completely separate.

Carefully pour the curd mixture through the cheesecloth (reserve the liquid, or whey, for another use). Allow the curds to drain through the lined colander into the bowl for about 20 minutes, or until the desired consistency is achieved.

After it has drained, use as desired or store refrigerated for 3 to 5 days.

Fresh Paneer

Paneer is an Indian curd cheese. It's made just like ricotta but finished differently. Use it in Saag Paneer (page 135).

Make the Whole-Milk Ricotta Cheese as directed on the previous page, but after you put the curds in the cheesecloth and before you leave them to drain, twist the cloth to squeeze out all the liquid. Tie the cheesecloth and allow the curds to drain through a colander into the bowl for another 10 minutes.

After it has drained, give the cheesecloth a few more good squeezes and lay it on a plate (curds still in the cheesecloth). Put a heavy pot on top of the cheese and press. Pick up the pot, and again, with the cheese still in the cloth, form the bundle into a square. Put the pot back on the wrapped cheese and let it sit for about 20 minutes. Use as desired or store refrigerated for 3 to 5 days.

Creamy Cashew Spread

This nut spread is a good non-dairy cheese alternative. The tasty plant-based option can be used as a substitute for milk-based cream cheese or ricotta.

Put all the ingredients in the bowl of a food processor and run for 2 minutes, or until the mixture becomes very smooth, like cream cheese. Store refrigerated in an airtight container for up to 7 days.

Makes about ¾ cup spread

1 cup raw cashews
¼ cup water
1 tablespoon fresh lemon juice
1 garlic clove, peeled
1 tablespoon white wine vinegar

Sun-Dried Tomato Pesto

The rich, concentrated flavor of sun-dried tomatoes is the star here. Use a spoonful in dips, spread on bread, or toss with pasta.

Makes about 1 cup pesto

1 ounce roasted almonds (about 23 nuts)

1 (8.5 ounce) jar sun-dried tomatoes packed in olive oil

1 tablespoon fresh lemon juice

1 tablespoon flat-leaf parsley leaves

8 to 10 fresh basil leaves

⅛ teaspoon kosher salt

⅛ teaspoon freshly ground black pepper

Put the almonds in the jar of a food processor and run for about 30 seconds, or until finely chopped. Add the tomatoes, including the olive oil, lemon juice, parsley, basil leaves, and salt and pepper. Pulse 6 to 8 times, or until the mixture becomes spreadable but has a slight texture remaining. Use immediately or store refrigerated in an airtight container for up to 2 weeks.

Hemp Seed Basil Pesto

I make triple batches of this and keep some in the freezer at all times. Hemp seeds replace nuts in this traditional pesto for a protein boost.

Makes about 1 ¼ cups pesto

3 cups packed fresh basil leaves

½ cup hemp seed hearts

¾ cup (about 2¼ ounces) coarsely grated Parmesan cheese

½ teaspoon kosher salt

¼ teaspoon freshly ground black pepper

½ cup garlic-infused olive oil, plus slightly more if needed

Put the basil, hemp seeds, Parmesan cheese, salt, and pepper in the bowl of a food processor and pulse several times until coarsely chopped. Add the olive oil and pulse again until combined and the desired texture is achieved. Use immediately or store refrigerated in an airtight container for up to a week.

Chickpea Veggie Sauce

Heat the olive oil in a skillet over medium-high heat. Add the garlic, onion, and carrots. Sauté for 8 to 10 minutes, or until the vegetables are softened. Add the tomatoes, red pepper, chickpeas, bay leaf, salt, and pepper. Lower the heat and simmer for 10 minutes.

Remove the pan from the heat, remove the bay leaf, and, using a ladle, transfer the mixture to a food processor or high-speed blender jar. Process until smooth, adding some of the cooking liquid (or vegetable stock) a little at a time, but only if needed to reach the desired consistency. Return the sauce to the pan and reheat. Stir in the oregano and chili flakes and additional salt or pepper, if needed. Use as desired or store refrigerated for up to 5 days or in the freezer for up to 3 months.

Makes about 3 cups sauce

2 tablespoons extra-virgin olive oil

2 garlic cloves, minced

1 small onion, sliced

2 carrots, peeled and diced

1 (28-ounce) can fire-roasted tomatoes

1 roasted red pepper, chopped

1 ½ cups cooked or canned chickpeas, drained, ¾ cup cooking liquid reserved

1 bay leaf

1 teaspoon kosher salt, plus more if needed

½ teaspoon freshly ground black pepper, plus more if needed

Vegetable stock (optional)

1 teaspoon fresh oregano leaves

1 teaspoon red pepper chili flakes

Cook's Note:

You may use jarred red peppers here, or see page 28 for how to roast them yourself.

Three-Grain Porridge

Cooking this combination of grains together (actually grains and seeds, as both quinoa and amaranth are technically seeds but treated like grains) hails from my friend Rebecca Shim, a former executive chef at the Menla Retreat in Phoenicia, NY. She used this nutrient-dense porridge, which she calls very forgiving, in a number of recipes. Heat it in the morning with a splash of milk and handful of fruit, or chill it in a pan to make polenta-like squares.

Put all the grains in a large pot with the water, vegetable stock, and salt. Bring to a boil. Lower the heat and simmer, uncovered, for 25 minutes, or until the millet is tender. (The other two grains cook more quickly, so they'll be done; it's the millet that needs to be tender but not mushy, as the slightest crunch should remain.)

Use as desired or spread evenly in a 9 x 13-inch pan lined with parchment overhanging on two sides. Use a spatula to press down the mixture into the pan. Cover with plastic wrap and chill for at least 30 minutes. After that time, cut the chilled porridge into 12 pieces. Fry the chilled squares in a skillet with a tablespoon of olive oil for about 5 minutes per side.

Use these squares as a polenta-like base, as in Crispy Three-Grain Cake with Mozzarella and Tomatoes (page 157). Keep refrigerated for up to 3 days.

Makes 8 cups porridge

¾ cup uncooked amaranth

¾ cup uncooked quinoa

¾ cup uncooked millet

6 cups water

2 cups vegetable stock, store-bought or homemade (page 35)

1½ teaspoons kosher salt

1 tablespoon extra-virgin olive oil (if frying as squares)

Breakfast

When the scent of banana muffins or sweet maple granola baking permeates our house, big smiles are guaranteed from my kids no matter how sleepy they may be. Busy weekdays don't usually allow for such luxuries, but we always find time for a quick smoothie or bowl of oats. Breakfast—comprising protein-rich options, such as eggs, tofu, whole grains, seeds, and nuts—is a meal we never skip. While instant oatmeal packs may be convenient, the processed oats are stripped of so much nutrition. Whole-grain breakfast porridges take time, but are worth every second. They satiate and make you feel full longer. I like to cook up a grain on Sunday nights so it's ready to go all week. If even a spoonful of nut butter on whole-grain toast, morning is the time I fuel up with protein to raise my brain's tyrosine levels so I feel awake, alert, and ready to take on the day.

Sweet Potato Almond Milk Smoothie

Creamy with a hint of spice, this smoothie tastes like liquid pie. It seems like a big treat but is chock-full of nutritional goodness with a nice dose of vitamins A and E. Hemp seeds are my superfood of choice for adding to smoothies because they've got more protein per ounce than almost any other seed. One of these for breakfast makes a happy start to the day.

Serves 1 | 7 g protein

⅓ cup roasted sweet potato, mashed

1 frozen banana, cut into several pieces

1 cup almond milk, store-bought or homemade (page 38)

1 tablespoon almond butter, store-bought or homemade (page 37)

½ teaspoon pure vanilla extract

¼ teaspoon ground cinnamon

⅛ teaspoon freshly grated nutmeg

1 tablespoon hemp seeds

Handful of ice

Cook's note:

Roast one or two sweet potatoes in advance (page 30) and freeze them in 1/3-cup portions. You can blend up the frozen sweet potato, but you may want to add more milk because it'll yield a thicker drink (kind of like ice cream!).

Put all the ingredients in a blender jar and process for 1 minute, or until smooth. Drink immediately.

Green Tea Pea Smoothie

Just like hemp seeds, chia seeds are a great addition to smoothies for an energy-rich break-fast. In addition to protein, you'll get a nice amount of fiber (6 grams) from this ingredient combo, which will make you feel full longer. Loaded up with immune-boosting vitamins, like Vitamin C from pineapple, and antioxidants from green tea, this grassy, fresh, not-too-sweet drink is refreshing and light.

Serves 2 | 5 g protein

1½ cups brewed green tea, chilled

½ apple, cored and chopped into large pieces

1 (1-inch) piece fresh ginger, peeled and grated

2 large kale leaves, stemmed, torn into pieces (about 1 cup)

½ cup frozen green peas

½ cup fresh pineapple (2-inch chunks), chopped

1 tablespoon chia seeds

1 tablespoon fresh lime juice

2 teaspoons honey, or more to taste

2 handfuls of ice

Put all the ingredients in a blender jar and process for 1 minute, or until smooth. Drink immediately.

Warming Breakfast Broth

A savory broth hits the spot many a morning. I immediately feel nourished from the warm, steamy liquid tinged a dark golden from healing turmeric, which can alleviate joint inflammation. In addition to the subtle taste of the medicinal spice, there's a slight hint of ginger, which invigorates, and tofu provides protein for energy.

Put the dashi in a small saucepan over medium-low heat. Add the turmeric and whisk until smooth. Add the carrot and bring to a simmer. Cook for 5 minutes, or just until the carrot is slightly softened. Lower the heat so the broth stays below a simmer. Ladle ½ cup or so of the broth into a mug and whisk in the miso paste until smooth. Pour the miso mixture back into the pan and stir it well. Continue to heat over medium-low heat. Add the tofu and ginger slices and heat until the tofu is heated through, about another minute or two.

Divide between two bowls and squeeze in a lemon slice. Drink immediately.

Serves 2 | 8 g protein

2 cups Vegetarian Shiitake Dashi (page 36)

1 teaspoon ground turmeric

1 small carrot, peeled and cut into ¼-inch dice

2 tablespoons white miso paste

1 (3-ounce) piece firm tofu, cut into ¼-inch dice

1 (½-inch) piece fresh ginger, peeled and thinly sliced into rounds

2 lemon slices

Maple Granola Clusters

To clump or not to clump, that is the question. I've found granola to be as polarizing as, say, chocolate chip cookies (some like 'em soft, some like 'em crispy). It seems some folks like large clusters they can pop in their mouths; others, a grainier, less clumpy mixture. Because of this, I've given you two ways to prepare this granola. You should also consider this recipe a base for your own taste preferences. You can swap in any nut, add dried fruit, replace the maple syrup with honey, or substitute olive oil for the coconut oil. Just keep the proportions the same and you're good to go.

Makes about 5 cups | 10 g protein in a ½-cup serving without yogurt

3 cups old-fashioned rolled oats
1½ cups unsweetened shredded coconut
1¼ cups chopped pecans
½ cup raw pumpkin seeds
½ cup raw sunflower seeds
¼ cup ground flaxseeds
½ cup pure maple syrup
2 tablespoons coconut oil, melted
¼ teaspoon kosher salt
2 large egg whites

Vanilla Bean Yogurt:
1 vanilla bean
3 tablespoons pure maple syrup per 2 cups whole-milk plain yogurt
Fresh berries

Cook's Note:

This recipe makes granola that clumps. If you prefer it flakier, reduce the egg whites to one, don't press the granola into the baking sheet, and stir the mixture several times while it's baking to prevent clusters from forming. For a vegan granola, leave out the egg whites altogether.

Preheat the oven to 325°F. Line two 11 x 17-inch rimmed baking sheets with parchment paper.

In a large bowl, stir together the oats, coconut, pecans, pumpkin and sunflower seeds, and ground flaxseeds. Drizzle the maple syrup and coconut oil over the mixture and sprinkle with the salt. Toss until well combined.

In a medium-size bowl, whisk the egg whites until foamy, about 2 minutes, and toss them with the mixture, ensuring everything is evenly coated.

Divide the granola between the two prepared pans and lightly press it down with the back of a spatula. Bake for 25 to 30 minutes, or until golden brown. Remove from the oven and allow it to cool. Then, break the granola into largish clumps and store in an airtight container for up to 3 weeks.

To make vanilla bean yogurt: Simply split the bean lengthwise, and using a sharp knife, scrape the black seeds into the yogurt. Stir in the maple syrup and serve with the granola and fresh berries.

Brown Rice Chia Seed Porridge

This hearty breakfast porridge is really like a warm rice pudding. Instead of cooking the rice fully with milk, though, you start with leftover rice (or you can always cook up a batch to use specifically for this porridge). If you're making rice for dinner one night, keep this recipe in mind and even make a little extra.

You can top this breakfast bowl with any seasonal fruit, but in spring and early summer when strawberries are at their sweetest and rhubarb is readily available, that combination is irresistible. If you don't have time to cook fruit, add a dollop of jam to your bowl instead. If you use cow's milk to make the porridge, the protein content will be greater than if you use nut milk. You can always sprinkle extra chia seeds directly on the porridge for a protein boost.

Make the strawberry-rhubarb compote: Put the strawberries and rhubarb in a small pan with the sugar and heat over medium-low heat. Stir frequently, and as you do, mash the mixture against the side of the pan with the back of the spoon. You'll notice after 10 minutes or so that liquid will start to develop in the pan. At this point, turn up the heat to medium so the mixture simmers, and cook for another 3 to 4 minutes. Add the chia seeds, stir well, and cook for another minute, then remove it from the heat and set it aside.

Meanwhile, make the creamy rice: Put the rice and 3 cups of the milk in a medium-size saucepan over medium heat. Stir to combine, then add the maple syrup, salt, and vanilla. When the mixture comes to a simmer, lower the heat to a very low simmer. Stirring the mixture often, cook until the rice is softened and the mixture is creamy, about 15 minutes. Once the milk is absorbed, add the remaining ½ cup of milk and bring to a simmer again, then just continue to cook it for a minute or two, or until the milk is heated through.

Divide the creamy rice among four bowls and top with the strawberry-rhubarb compote. Serve immediately.

Serves 4 | 14 g protein with cow's milk

For the strawberry-rhubarb compote:
- 1 pint strawberries (about 2 cups sliced)
- 4 large rhubarb stalks, cut into 1-inch pieces (about 2 cups)
- 2 tablespoons cane sugar, or more to taste
- 3 tablespoons chia seeds

For the creamy rice:
- 4 cups cooked short-grain brown rice
- 3½ cups milk of your choice, divided
- 3 to 4 tablespoons pure maple syrup, or more to taste
- ¼ teaspoon kosher salt
- 1 teaspoon pure vanilla extract

Cook's Note:

If rhubarb isn't available, substitute all strawberries, fresh or frozen, and throw in 1 teaspoon of chopped fresh tarragon or mint. Also, you can substitute different types of rice, such as white and/or basmati.

Overnight Steel-Cut Oats

While steel-cut oats are nearly identical to rolled oats in their nutritional profile, including the same grams of protein per cup, steel-cut oats are less processed, which causes them to be digested more slowly and have a lower glycemic index. Just as their name states, steel-cut oats are oat kernels that have been cut or chopped, as opposed to rolled oats, which have been steamed, rolled, and toasted. The difference is slight and both forms of oats are extremely good for you. Steel-cut oats, however, have a slightly chewy, more robust texture that some find more satisfying than that of traditional rolled oats. The downside to steel-cut oats is that they take significantly longer to cook, which is why this recipe calls for a slow cooker. Put in the oats before you go to bed and you'll wake to a pot of warm deliciousness—especially gratifying on a cold winter's morn.

This recipe makes a good batch of basic oats boosted with flaxseed. The secret to a good bowl of oatmeal is in the toppings.

**Serves 4 | 12 g protein
with cow's milk**

1 teaspoon coconut oil, melted, or
    unsalted butter, at room temperature

1 cup steel-cut oats

1¾ cups milk of your choice

¼ teaspoon kosher salt

3 cups water

2 tablespoons ground flaxseeds

2 tablespoons pure maple syrup

Fruit, nuts, and seeds, for topping (see
    cook's note)

Extra milk of choice or half-and-half, for
    serving

Cook's Note:

For toppings, try a spoonful of jam, pecans, and dried coconut, a dollop of yogurt with cinnamon and honey, or even pumpkin seeds and dried cherries.

Coat the inside of a slow cooker with the coconut oil or butter. Add all the other ingredients, except the toppings, to the slow cooker, give it a stir or two, and set it to LOW. Allow the oatmeal to cook for 6 or 7 hours. Be careful not to leave it for too long, though, as it will start to stick to the sides and burn. When ready to serve, scoop the oatmeal into bowls and add your desired toppings. I always add a dash of milk or half-and-half, too.

Creamy Amaranth Banana Porridge

Did you ever eat farina as a kid? A smooth, warm wheat porridge, it was one of my favorite childhood breakfasts. While it tasted good, little did I know that much of the cereal's fiber and goodness was lost when the wheat was ground and sifted. Recently I've discovered amaranth—actually a seed and not a grain. It yields a creamy consistency very similar to that of farina when cooked but, boy, does it pack a nutritional punch: Not only does amaranth have a complete protein profile, it has double the protein and a whopping five times more fiber than farina. It is also naturally high in iron and lysine, an essential amino acid that most wheat grains lack. It does take longer to cook, though, but it's certainly worth every extra minute. This breakfast is a terrific vegan and gluten-free option.

Melt the butter or oil in a medium-size saucepan over medium heat. Add the banana slices and sauté for several minutes, stirring continuously until they start to soften and brown slightly in spots. Sprinkle with the cinnamon, nutmeg, and salt and stir to combine.

Stir in the water, 1¼ cups of the almond milk, and the maple syrup and almond butter. Give a few good stirs to mix everything together, then add the amaranth. Bring the mixture to a boil, lower the heat to a simmer, and cover. Stirring every 5 minutes to ensure the amaranth doesn't stick to the pot, cook for 30 to 35 minutes, or until the porridge is creamy and the seed is softened.

Finally, stir in the remaining ½ cup of almond milk and heat through. Divide between two bowls and top with a sprinkling of chopped almonds.

Serves 2 | 15 g protein

1 teaspoon unsalted butter or coconut oil

1 ripe banana, sliced

¾ teaspoon ground cinnamon plus more for topping

¼ teaspoon freshly grated nutmeg

Kosher salt

1 cup plus 2 tablespoons water

1¾ cups almond milk, divided

1 to 2 teaspoons pure maple syrup, plus more for drizzling

2 tablespoons almond butter

¾ cup uncooked amaranth

Handful of chopped almonds, for topping

Cook's Note:

Again, like the other porridges and warm grain cereals in this chapter, use amaranth as a base for your favorite seasonal fruits, nuts, and seeds. Here I cooked bananas before adding the amaranth, but you can just cook the grain without any additions and top it with whatever you have on hand.

Breakfast Barley with Raspberries and Honey Figs

Nutty, chewy barley is a superb alternative to the standard oatmeal breakfast routine. Rich with manganese and selenium, barley has half the glycemic index of wheat. It's also extremely high in fiber, iron, and calcium. Sounds pretty great, huh? It is, but it does take a long time to cook. Perhaps that's why we don't see it as often in porridge as oats or wheat.

When buying barley, you may find any or all of three types: hulled, pearled, or hull-less. Hulled barley, also known as barley groats, is the whole barley grain with the outermost hull removed. Pearl barley is hulled barley with the hull and outer bran layer mechanically removed (pearled or polished) and is no longer considered a whole grain. (Hulled and pearled barley are akin, respectively, to brown and white rice.) Then there's a third kind of barley— hull-less—a variety grown with a hull so loosely attached that it comes off during harvesting.

It's much more common to find hulled or pearl barley, which, even though it is less whole, is still a very good, heart-healthy grain choice. Pearl barley will also yield a creamier result when cooking.

Serves 4 | 10 g protein with cow's milk

1 cup hulled, pearl, or hull-less barley

6 cups water, divided, plus more if needed

½ teaspoon kosher salt

½ pint (8 ounces) raspberries

1 pint whole figs (about 10)

2 tablespoons honey, plus more for serving

¼ cup chopped hazelnuts

1 cup milk of your choice

Raspberries, for serving

Cook's Note:

The cooking times and water amounts, which will depend on the type of barley you use, are approximate. Be aware that hull-less barley may take over an hour to cook.

Soak the barley in 2 cups of the water overnight. The next morning, drain the barley and discard the water. Put the soaked barley and remaining 4 cups of water in a medium-size saucepan over medium-high heat.

Bring the water, barley, and salt to a gentle boil, stir to prevent any sticking, and lower the heat so the mixture comes to a simmer. Cover and cook for 45 minutes. After that time, check the grains for doneness. They should be tender but a teeny bit chewy. Cook longer, if necessary. If you continue to cook the grains but all the water is absorbed, add a little bit more water. If the grains are tender but some water remains, you can either drain it off or allow them to sit, covered, for 10 minutes and the cooked barley should absorb the water. (It's pretty hard to overcook barley.)

CONTINUED

While you are cooking the grains, preheat the oven to 375°F. Place the figs in a small ovenproof dish and drizzle with the honey. Put into the oven and bake for 10 minutes.

Next, toast the hazelnuts. Put them in a dry skillet over medium-high heat and stir constantly for 3 to 5 minutes, or until they start to smell toasted and nutty.

When the grains are finished cooking, add the milk and stir well, continuing to heat over low heat until the grains become porridgelike. Top with fresh raspberries, the figs and hazelnuts, and a drizzle of honey. Serve immediately.

Black Bean Breakfast Burrito

I never tire of breakfast burritos because they can always be changed up. I use burritos as a whatever's-in-the-fridge catchall. Here I used leftover roasted butternut squash. Roasted red peppers, sautéed zucchini, and even eggplant work well and may be substituted for the squash. Get creative. Try different spreads (sriracha mayo perhaps?) or cheeses (how about smoked mozzarella?). Whatever your preference, use this recipe as a base and make it your own.

Preheat the oven to 350°F. Stack the four tortillas and wrap them in foil. Set aside.

Heat 2 teaspoons of the olive oil in a sauté pan over medium-high heat and add the onion. Cook until softened, stirring frequently, about 5 minutes. Add the beans, green chiles, roasted squash (or other vegetable) and ½ teaspoon of the salt and cook for another few minutes, until heated through. Scrape the mixture into a bowl, cover to keep warm, and set aside.

Put the tortillas in the preheated oven.

Crack the eggs into a bowl and whisk for a minute. Next, add the milk and the remaining ½ teaspoon of salt and continue to whisk for 1 more minute. In the same sauté pan used for the bean mixture, heat the remaining 2 teaspoons of olive oil. Pour the whisked eggs into the pan and let them sit for about 15 seconds. Then, using a wooden spatula or spoon, gently scrape the eggs from the pan and fold them while carefully stirring them. Continue to stir until the eggs are just set. Remove from the heat.

Serves 4 | 23 g protein

- 4 (10-inch) soft whole-wheat flour tortilla shells
- 4 teaspoons extra-virgin olive oil, divided
- ½ medium yellow onion, chopped (about ¾ cup)
- 1½ cups cooked or canned black beans, drained
- 1 (4-ounce) can green chiles, diced
- ½ cup butternut squash (or other cooked vegetable), roasted and diced
- 1 teaspoon kosher salt, divided
- 4 large eggs
- 3 tablespoons whole milk

For assembly:
- 1 pint grape tomatoes, cut in half
- ¼ cup fresh cilantro, chopped
- 3½ ounces Monterey Jack cheese, grated
- 1 avocado, sliced
- 2 tablespoons sour cream (optional)

CONTINUED

To assemble: Arrange your toppings so they are accessible and remove the tortilla shells from the oven. Lay one tortilla flat and scoop one-quarter of the scrambled egg mixture into the burrito. Add one-quarter of the bean mixture and top with tomatoes, cilantro, cheese, avocado, and sour cream, if desired. To roll your burrito, fold the left and right edges of the shell over your fillings, then, starting from the top edge, roll the filled tortilla down toward the bottom edge. You should have a cylinder. Repeat to make the remaining burritos. Cut the burritos in half and serve. You may want to wrap them in foil to keep them secure when eating—it also works well for reheating them.

Cook's Note:

You may freeze burritos for a quick weekday breakfast. Simply wrap extra burritos (omitting the avocado and sour cream) in plastic wrap and then foil. Store the burritos in a resealable plastic bag or plastic container in the freezer. When ready to eat, unwrap the burrito, discarding the plastic wrap, and rewrap with the foil. Warm the frozen burrito in a 400°F oven for 30 to 40 minutes. You may also heat a frozen burrito in the microwave. Unwrap the plastic and foil and cover with a wet paper towel. Heat on HIGH for about 1½ minutes.

Carrot Millet Griddlecakes with Summer Peaches

What these griddlecakes are not is a traditional fluffy pancake. Instead, they are tangy, eggy, and quite thin (although not quite crepe thin). I find these characteristics quite appealing here. Millet flour, which is rich in amino acids, especially methionine, adds a sweetness to the batter. But beware: Ground millet spoils quickly and can impart a bitter flavor into baked goods, so be sure yours is fresh—I've learned the hard way!

Place the peaches in a small saucepan with the water and 2 tablespoons brown sugar. Bring to a simmer over medium heat. Lower the heat and continue to cook, stirring frequently, for 30 minutes, or until the mixture is reduced and becomes syrupy. If it's not thickening, increase the heat slightly and continue to stir.

Meanwhile, whisk the flours, baking powder, remaining 2 tablespoons brown sugar, salt, ginger, and nutmeg together in a bowl. In a smaller bowl, whisk together the buttermilk and egg. Add the carrot and almond extract, if desired, and stir. Make a well in the dry ingredients and pour in the wet. Stir until just combined.

Melt a couple teaspoons of butter or coconut oil on a griddle over medium heat. Pour a scoop of batter (about 2 tablespoonfuls) onto the heated surface to make a 4-inch round. Repeat. Cook until small bubbles appear in the surface of the batter and start to burst uniformly over the pancake. Flip the pancakes and cook for several minutes on the other side. When golden brown on both sides, transfer to a serving platter. Continue to cook the pancakes until all the batter is used, adding more butter or oil to the pan as needed.

Serve with butter, maple syrup, the warm peaches, and a dollop of yogurt, if desired.

Makes 16 (4-inch) pancakes | 12 g protein

- 1½ pounds peaches (about 4 peaches), pitted and cut into chunks
- ¼ cup water
- 4 tablespoons dark brown sugar, divided
- 1 cup millet flour or ground millet
- ½ cup white whole-wheat flour
- 1 tablespoon baking powder
- ½ teaspoon kosher salt
- ¼ teaspoon ground ginger
- ¼ teaspoon freshly grated nutmeg
- 2 cups buttermilk
- 1 large egg
- ½ cup finely shredded carrot (about 2 large carrots)
- ⅛ teaspoon almond extract (optional)
- 1 to 2 tablespoons unsalted butter or coconut oil, for cooking
- Butter, pure maple syrup, and plain or vanilla yogurt, for serving (optional)

Quinoa Oat Breakfast Cookies

Pack a couple of these fiber-filled cookies for an easy breakfast on the go. They are somewhat soft, though the nuts and coconut give them a good crunch, and they make a good replacement for a muffin. And just because they're called cookies, don't expect them to be too sugary. They're not a dessert cookie. However, the dried cherries add a natural sweetness. Anyway, if you're going to eat cookies for breakfast, why not make them relatively healthy with some wholesome grains?

Make 32 cookies | 8 g protein in 2 cookies

½ cup (1 stick, or 4 ounces) unsalted butter, at room temperature, or coconut oil, melted

¼ cup crunchy almond butter

⅔ cup pure maple syrup

1 large egg

1 teaspoon pure vanilla extract

1 cup oat flour

1 cup toasted wheat germ

1 teaspoon kosher salt

1 cup cooked quinoa

½ cup old-fashioned rolled oats

1 cup dried cherries

½ cup almonds, chopped

½ cup unsweetened shredded coconut

Preheat the oven to 375°F. Line two baking sheets with parchment paper.

In the bowl of a stand mixer fitted with the paddle attachment, beat the butters together until fluffy. Add the maple syrup, egg, and vanilla and continue to beat until thoroughly combined.

Add the oat flour, wheat germ, and salt and beat for another 10 to 15 seconds. Add the quinoa and oats and beat until just combined. Stir in the cherries, almonds, and coconut.

Using a tablespoon measure, scoop mounds of dough about 2 inches apart onto the prepared baking sheet. Bake for 13 to 15 minutes, or until a toothpick inserted into the center of a cookie comes out clean. Remove from the oven and allow to cool for 5 minutes on the baking sheet before transferring to a wire rack.

Banana Buckwheat Muffins

Pair these snack-size muffins with a smoothie for a perfectly portable breakfast. The balance of buckwheat and white whole-wheat flour gives this baked good a wonderful texture, kind of nutty but still tender. Flaxseed, a quality source of protein and fiber, is used ground here, as opposed to whole, to make the seeds' health benefits more available. Eat them warm with jam straight from the oven and you'll want them as much for their taste as for their nourishment.

Preheat the oven to 350°F. Grease a 24-cup mini muffin tin with baking spray (or lightly coat with oil) and set aside.

Combine the bananas, oil, and sugar in the bowl of a stand mixer fitted with the paddle attachment. Mix for 30 seconds on medium speed, or until thoroughly combined, then add the eggs, yogurt, and vanilla. Run the mixer again for 15 seconds or so, until the ingredients are well combined.

Add the dry ingredients at once and then run the mixer on low speed to incorporate everything and mix until just combined.

Using a tablespoon or scoop measure, fill the prepared muffin cups ¾ full. If desired, place one slice of banana on top of the batter in each muffin cup. Bake for 12 minutes, or until a toothpick inserted into the center of a muffin comes out clean. Let the muffins cool in the pan slightly before removing.

Makes 24 mini muffins | 6 g protein in 3 mini muffins

Baking spray or oil, for pan
2 ripe bananas, mashed
¼ cup coconut oil, melted, or canola oil
½ cup coconut sugar or cane sugar
2 large eggs
½ cup whole-milk plain yogurt
½ teaspoon pure vanilla extract
¾ cup light buckwheat flour
½ cup white whole-wheat flour
¼ cup ground flaxseeds
2 teaspoons baking powder
½ teaspoon kosher salt
½ teaspoon ground cinnamon
¼ teaspoon freshly grated nutmeg
24 slices banana, for topping (optional)

Cook's Note:

Look for light buckwheat flour, such as Bouchard Family Farms Acadian Light Buckwheat Flour. Substituting dark buckwheat flour will change the texture, making these muffins much denser.

Savory Cheddar Pinto Bean Muffins

These dense, savory corn muffins work for a grab-on-the-go kind of morning, or you can savor them with an egg frittata or generous arugula salad for brunch. There are no substitutes here for the cheese, a must for the rich moist crumb.

Makes 12 muffins | 9 g protein in 1 muffin

Baking spray or oil, for pan
1 cup white whole-wheat flour
¾ cup cornmeal
1½ teaspoons baking powder
½ teaspoon baking soda
1 teaspoon kosher salt, divided
⅛ teaspoon cayenne pepper
¼ cup plus 1 tablespoon extra-virgin olive oil, divided
½ cup scallions, finely chopped
1 garlic clove, minced
1 fresh jalapeño pepper, seeded and finely diced
1½ cups cooked or canned pinto beans, drained and rinsed
1¼ cups buttermilk
1 large egg
¼ cup fresh cilantro leaves
½ roasted red pepper, finely diced (about ½ cup)
1 cup grated sharp Cheddar cheese, divided

Cook's Note:

You may either use jarred roasted red pepper or roast your own (page 28). If you use jarred, be sure to drain the peppers well and pat them lightly with a paper towel.

Preheat the oven to 350°F. Grease a 12-cup muffin tin with baking spray or brush with oil.

In a medium-size bowl, stir together the flour, cornmeal, baking powder, baking soda, salt, and cayenne. Set aside.

Heat 1 tablespoon of the olive oil in a skillet over medium heat. Sauté the scallions for about 2 minutes, add the garlic and jalapeño, and continue to cook for another 3 minutes. Stir in the beans and toss gently to combine. Sprinkle with ½ teaspoon of salt. Set aside.

In another bowl, whisk together the buttermilk, egg, and the remaining ¼ cup of olive oil. Make a well in the dry ingredients. Stir in the milk mixture, then gently stir in the bean mixture, cilantro, red pepper, and ½ cup of grated cheese.

Fill the prepared muffin cups ¾ full with batter. Top with the remaining ½ cup of grated cheese. Bake for 25 minutes, or until golden. Let the muffins cool in the pan slightly before removing.

Root Vegetable Hash with Fried Eggs

Many things are made better when you top them with an egg. Mind you, the egg mustn't be overcooked. A dry, crumbly center just won't do. A deep yellow, runny, exquisite yolk oozing over this potato turnip hash—now, that's downright decadence. I've used kohlrabi in place of the turnip, and added squash at times, too. While the vegetables are savory and full of flavor, let's be honest: In this case, it really is all about the egg.

Serves 4 | 13 g protein

1 large sweet potato (about 1 pound), peeled and cut into ½-inch dice

1 large Idaho potato (about 1 pound), peeled and cut into ½-inch dice

1 medium turnip (about ½ pound), peeled and cut into ½-inch dice

3 tablespoons plus 2 teaspoons extra-virgin olive oil, divided, plus more if needed

1 teaspoon kosher salt, plus more to taste, divided

1 red onion, peeled and chopped (about 1 heaping cup)

1 large garlic clove, peeled and minced

1 tablespoon rosemary leaves, chopped

4 large eggs

Freshly ground black pepper

2 tablespoons fresh parsley, chopped

Cook's Note:

The key here to a crispy, slightly caramelized-on-the-edges kind of hash is to stir as little as possible. Don't be stingy with the oil as it will help to crisp up the veggies, and try to use a spatula to flip the hash instead of stirring it.

Preheat the oven to 400°F. Put the diced potatoes (both sweet and Idaho) and turnip on a rimmed baking sheet. Toss with 1 tablespoon olive oil, ½ teaspoon salt, and freshly ground pepper. Roast for 15 minutes or until softened, tossing the vegetables once with a spatula halfway through the cooking time.

Meanwhile, heat 2 more tablespoons of the olive oil in a large skillet over medium heat. Add the onion and sauté for 8 minutes, or until softened, stirring frequently. Add the garlic and rosemary, stir well, and cook for 1 more minute. Add the cooked potatoes and the turnip and toss to combine. Spread the mixture evenly into the skillet and leave it to cook for 2 to 3 minutes. Turn gently with a spatula. Add a little more oil if the vegetables look dry. Lower the heat and cook for another 5 to 8 minutes, or until cooked through and a crispy edge starts to form on the vegetables. Remove from the heat and divide the mixture among four plates.

Return the pan to the stovetop and heat the remaining 2 teaspoons oil over medium heat. Crack the eggs into the pan and fry sunny-side up. You may put a lid on the pan to ensure even cooking for a minute or two. You want to remove the eggs from the pan when the yolks are still runny but the whites are completely set. Place the eggs on top of your hash. Sprinkle with salt, pepper, and parsley and serve.

Classic Egg and Cheese Sandwich

O, egg sandwich, how I love thee. My love affair with a perfectly cooked egg (runny yolk, thoroughly cooked whites, a teeny bit crispy on the edges) on some type of warm, delicious, buttery bread with melted cheese is a long-standing one. The humble egg and cheese on a roll is invariably satisfying. However, this dependable sandwich can also get dressed up—and may just surprise you. Try a fried tomato, avocado slice, and sautéed spinach on whole-grain ciabatta and you're likely to fall in love, too.

Preheat the oven to 350°F. Line a rimmed baking sheet with parchment paper. Lightly butter the rolls and place them flat side up on the prepared baking sheet. Top with the cheese slices. Place the rolls in the oven for about 5 minutes to melt the cheese.

Meanwhile, heat 2 teaspoons of the olive oil in a medium-size skillet over medium heat. Sauté the garlic for about 30 seconds, then add the spinach, sprinkle with a pinch or two of salt, and sauté until wilted, 1 or 2 minutes. Remove the spinach from the pan but keep the heat on. Next, lightly fry the tomato slices in the pan for about 1 minute on each side, again sprinkling each with salt and pepper. Remove them from the pan. Check your rolls if you haven't done so already, and pull them out of the oven so you can slide a tomato slice directly onto each of the bottom rolls.

Add the remaining 2 teaspoons olive oil to the pan and continue to heat over medium heat. Crack the eggs into the pan and fry sunny-side up until the whites start to set, about 2 minutes. Sprinkle with salt and pepper. Remove the eggs from the pan when the yolks are runny but whites are completely cooked, about another minute or so.

To assemble your sandwiches, lay a slice or two of avocado on top of the tomato, then add an egg and top with a spoonful of sautéed spinach and the top bun. Serve immediately.

Serves 4 | 25 g protein

4 whole-grain ciabatta rolls or other hearty rolls, cut in half

1 tablespoon unsalted butter, at room temperature

4 ounces white Cheddar cheese, thickly sliced

4 teaspoons extra-virgin olive oil, divided

1 garlic clove, minced

1 bunch fresh spinach, washed and coarsely chopped

Kosher salt

1 large tomato, sliced

Freshly ground black pepper

4 large eggs

1 avocado, sliced

Lentil, Spinach, and Tomato Frittata

Spinach, tomatoes, and goat cheese are a cherished trinity of mine. Pile them high on a piece of crusty bread, toss them with pasta, or as I often do, include them in a frittata. The marriage of these three flavors with eggs and lentils is a solid one. While lentils are a rich source of protein, they're also high in fiber, iron, and folate.

Serves 6 | 20 g protein

½ cup dried green lentils, rinsed

1 bay leaf

1 garlic clove, peeled

½ red onion, cut into half-circles (about ¾ cup)

2 tablespoons extra-virgin olive oil

1 pint grape or cherry tomatoes, cut in half

½ teaspoon kosher salt, divided

3 cups (about 3 ounces) baby spinach leaves

⅛ teaspoon freshly ground black pepper, divided

12 large eggs

½ cup whole milk

1 cup fresh basil leaves, cut into 1-inch-wide strips

3 ounces goat cheese

Preheat the oven to 350°F.

Put the lentils, bay leaf, and garlic clove in a saucepan and cover with 2 inches of water. Bring to a boil. Simmer over medium heat for about 30 minutes, or until the lentils are tender. Remove from the heat and set aside.

Meanwhile, in a 12-inch cast-iron or ovenproof skillet, sauté the onion in the oil for 4 to 5 minutes, or until softened. Add the tomatoes, season with ¼ teaspoon of the salt, and continue to sauté for another 3 minutes. Then, add the spinach, another pinch of salt, and the pepper, and continue to cook for another minute, or until the spinach is just wilted. Remove the garlic clove and bay leaf from the cooked lentils, drain any excess water, and add the lentils to the skillet. Toss to combine.

Meanwhile, whisk the eggs and milk together until frothy. Stir in the basil, remaining ¼ teaspoon of salt, and remaining pepper.

Lower the heat and add the eggs to the pan. Cook until the edges are just set, about 4 minutes. Using a tablespoon, dollop the goat cheese evenly across the top of frittata.

Place the skillet in the oven and bake for 20 minutes, or until just set in the center (it will puff slightly). Remove the frittata from the oven, cut into eighths, and serve warm or at room temperature.

Tarragon Egg Salad

For some reason, egg salad has never appealed to me much—it seemed kind of boring. It was always there—in deli cases and my mother's fridge—but I passed it over every time. Then I found myself one day with an enormous amount of boiled eggs. Egg salad was the reasonable answer. I doctored up the classic version with some tarragon and red onion. Voilà! I was converted. It makes an irresistible open-face sandwich with warm, homemade rye bread.

Place the eggs in a single layer on the bottom of a large pot and fill it with enough water to cover them by 2 inches. Place the pot over high heat, bring the water to a boil, remove the pot from the heat, and cover. Allow the eggs to sit in the water for 13 minutes. Meanwhile, prepare an ice bath. After the eggs have rested in the water, drain them and place them in the icy water for 5 minutes. Drain them again and set aside.

In a small bowl, whisk together the mayonnaise, mustard, vinegar, and salt.

Peel and chop the eggs and place them in a medium-size bowl. Add the celery, chives, capers, tarragon, and onion. Spoon the mayonnaise mixture over the eggs and toss gently with a spoon until everything is covered with the dressing. Crack pepper over the mixture. Serve on the rye bread and top with sliced radishes, if desired.

Serves 6 | 13 g protein

12 large eggs

3 tablespoons mayonnaise, store-bought or homemade (page 40)

1 tablespoon Dijon mustard

2 teaspoons cider vinegar

½ teaspoon kosher salt

1 celery stalk, diced

2 tablespoons chopped chives, plus more for garnish

1 tablespoon capers, chopped

1 tablespoon fresh tarragon leaves, chopped

2 tablespoons diced red onion, plus more for garnish

Freshly ground black pepper

A few thinly sliced red radishes, for garnish

No-Knead Dark Rye Bread (page 82)

No-Knead Dark Rye Bread

Once I learned about Jim Lahey's No-Knead Bread method, made famous by Mark Bittman in the New York Times, I was obsessed. Just a few minutes of effort yields a hot, crusty homemade loaf of bread that is chewy and airy with a golden, crispy crust.

Although Lahey's original recipe uses white flour, I've adapted his method to make a rye loaf. It's not really German rye bread, nor Russian black bread. It's a mix of white whole-wheat and rye flour that's made dark by cocoa and molasses, not the actual flour, and finished with a salty, seedy topping. It's well worth the little effort that's needed to make this bread when you smell it baking. Try toasting a piece with a big smear of butter, too.

Makes 1 (9-inch) loaf | 9 g protein in 2 slices

2¼ cups white whole-wheat flour

1 cup rye flour

2 tablespoons unsweetened cocoa powder

1 (¼-ounce [2¼-teaspoon]) packet dry instant yeast

1 ½ teaspoons kosher salt

1 cup water

½ cup beer

2 tablespoons molasses

1 tablespoon cider vinegar

For the topping:

2 teaspoons sunflower seeds

2 teaspoons caraway seeds

2 teaspoons black sesame seeds

2 teaspoons Maldon sea salt

In a large bowl, whisk together the flours, cocoa powder, yeast, and salt. Add the water, beer, molasses, and vinegar to the dry ingredients and stir well, ensuring all the ingredients are moistened. The batter will be sticky. Cover the bowl with plastic wrap and let sit for 4 hours in a warm place

To make the topping: Stir together the seeds and salt in a small bowl.

After the 4 hours have elapsed, put a Dutch oven in the oven and preheat the oven to 450°F. Meanwhile, turn out the dough onto a piece of parchment paper that will fit in the bottom of the Dutch oven. Form the dough into a round and sprinkle with the topping mixture, lightly pressing it into dough. Use a knife to make a slit or two in the top of the loaf. Transfer the dough along with its parchment into the heated Dutch oven. Cover it with a lid and bake for 30 minutes. Uncover the loaf and bake for another 30 minutes, or until the loaf is browned on top (which is hard to gauge because the dough is dark, but if you tap it with your finger the crust should feel almost crisp). Remove from the oven and allow it to cool slightly before removing it from the pan.

Kale and Carrot Tofu Scramble

I was a little late to the scrambled tofu game. It wasn't until Ankita, my lab partner in a nutrition course, introduced me that I discovered the many wonders of this versatile bean curd protein. Not only is tofu cholesterol- and gluten-free, it's an excellent source of amino acids and micronutrients. Plus, folks, it's really cheap, even the organic variety (at least relative to grass-fed meats). Pair crumbled tofu, made golden yellow from a spoonful of turmeric, with most any vegetable. Here it's curly kale and bright carrots, and you've got a nutritious, delicious meatless meal. This recipe, a favorite of mine, is adapted from the scramble Ankita developed for our lab.

Place the pumpkin seeds in a dry skillet and cook over medium heat for 2 to 3 minutes while stirring constantly. They will start to pop and puff up as they toast. Transfer them to a small plate or bowl and set aside.

Heat the olive oil in the same skillet and add the onion. Sauté for 5 minutes, or until softened. Next, add the carrot sticks and continue to sauté for about 3 minutes, or until the carrots are somewhat cooked but still crunchy. Add the crumbled tofu and mix well with the vegetables.

In a small bowl, whisk the mustard, ginger, turmeric, tamari, and water until smooth. Pour this mixture into the pan and stir to combine, coating the carrots and tofu. Next, add the kale leaves and stir-fry until the greens are slightly softened but not completely wilted. Sprinkle the entire mixture with kosher salt and black pepper, to taste. Divide the scramble among four plates and sprinkle with the pumpkin seeds and fresh cilantro leaves, if desired.

Serves 4 | 22 g protein

- 2 tablespoons raw pumpkin seeds
- 1 tablespoon extra-virgin olive oil
- 1 yellow onion, peeled and thinly sliced (about 1 heaping cup)
- 2 medium carrots, peeled and cut into 2-inch matchsticks
- 1 (15-ounce) package of organic firm tofu, drained and crumbled
- 1 tablespoon grainy Dijon mustard
- 1 teaspoon freshly grated ginger
- 1 teaspoon ground turmeric
- 2 teaspoons tamari
- 2 tablespoons water
- 4 to 5 curly kale leaves, stemmed, cut into bite-size pieces (about 1½ heaping cups)
- ½ to ¾ teaspoon kosher salt
- Freshly ground black pepper
- Fresh cilantro leaves for garnish, if desired

Lunch

While I never, ever skip breakfast, I admit that lunch is the meal I sometimes pass right by. I get so caught up in the day that come late afternoon, I grab whatever's in sight to devour. Not good behavior, I know. In fact, I'm embarrassed to admit that this happens, but that's why this chapter is filled with lunchtime favorites. They are enticing, flavorful meals that are meant to satisfy. A tofu bánh mì or a grilled vegetable and ricotta sandwich will always lure me to eat.

Many of these recipes take some planning, and are generous servings, but they're also dishes you can make on a Monday and eat throughout the week. And just because these recipes are in the lunch chapter doesn't mean they don't work for dinner. They are meals that will supercharge you anytime of the day.

Kale, Roasted Beet, and Edamame Salad

The equation here of robust kale leaves + satiating sweet roasted beets + steamed edamame + salted pistachios = a hearty salad that leaves you content without feeling weighed down. While the protein here isn't off the charts, it's still respectable, and the kale and edamame in the salad deliver tremendous amounts of vitamins A, C, and K.

Preheat the oven to 400°F. Spread the diced beets evenly on a baking sheet. Sprinkle the beets with the olive oil and salt, toss well, and spread out again evenly on the pan. Place in the preheated oven and roast for 20 to 25 minutes, tossing once if necessary. When the beets are tender, remove from the oven and set aside.

Meanwhile, make the dressing: Combine the ingredients for the dressing in a small jar with a lid and shake or stir with a fork until combined. Set aside.

To assemble the salad, cut the kale leaves into bite-size pieces and place in a large salad bowl. Pour about ¾ of the dressing on the leaves and start to massage them, rubbing them together. After a few minutes, you will feel the leaves soften and even notice that they're a brighter green. When you've gotten to this point, add the roasted beet, edamame, scallions, and pistachios. Add the rest of the dressing and toss. Season with more salt and pepper if needed.

Serves 4 | 12 g protein

3 beets, washed and peeled, cut into 1-inch dice
1 tablespoon extra-virgin olive oil
¼ teaspoon kosher salt
1 medium bunch curly kale, rinsed, dried, and stemmed
1 cup shelled edamame, steamed
4 scallions, white part only, chopped
1 cup shelled pistachios, roughly chopped

For the dressing:
⅓ cup extra-virgin olive oil
1 tablespoon balsamic vinegar
1 tablespoon tamari
1 teaspoon Dijon mustard
1 teaspoon honey
1 garlic clove, peeled and minced
1 tablespoon minced shallot
½ teaspoon kosher salt
¼ teaspoon freshly ground black pepper

Farro, Baby Beet, and Pea Shoot Salad

The beauty of this salad is in the simplicity of spring flavors—young peas, delicate pea shoots, and sweet baby beets. Farro, a fiber-rich grain, provides a chewy, nutty base for these tender vegetables. Paired with peas, this grain creates a complete source of protein. For non-vegans, add a piece of dreamy, creamy burrata for instant decadence.

Serves 6 | 19 g protein

For the dressing:
2 tablespoons red onion, finely diced
3 tablespoons extra-virgin olive oil
1 tablespoon fresh orange juice
1 tablespoon fresh lemon juice
2 teaspoons white wine vinegar
1 teaspoon honey
½ teaspoon kosher salt

For the farro salad:
1 cup uncooked farro (3 cups cooked)
2¾ cups water
1 bay leaf
¾ teaspoon kosher salt, divided
1 bunch beets (about ¾ pound),
    trimmed, peeled, and cut into
    1½-inch wedges
½ cups fresh peas
⅓ cup walnuts
Freshly ground black pepper
1 bunch radishes (about ½ pound),
    peeled and thinly sliced
2 to 3 cups pea shoots
1 pound burrata cheese (optional)

Cook's Note:

If pea shoots aren't available, you can substitute the same amount of arugula or other sprout or green.

Make the dressing: Combine all the ingredients in a jar or lidded container. Shake well and allow the flavors to meld while you prepare the salad.

Start the farro salad: Cook the farro by putting it in a medium-size saucepan with the water, bay leaf, and ½ teaspoon of the salt. Bring the water to a boil over high heat, then lower the heat to a simmer, cover, and cook for 30 minutes, or until the grain is tender yet still somewhat chewy. Drain the farro if any liquid remains and transfer it to a salad bowl.

Place a vegetable steamer basket in the saucepan and fill so the water is just below the basket. Put the beets in the steamer basket, put on a lid, and heat over medium-high heat until the water comes to a boil. Lower the heat to a gentle simmer and continue to steam the beets for about 15 minutes, or until fork-tender. When they are done, use a fork or pair of tongs to remove them from the steamer basket and set them aside to cool. You can then put the peas in the steamer basket for a few minutes to steam them until just tender. Add a little more water to the pan if necessary.

CONTINUED

While the beets are cooking, toast the walnuts in a small, dry skillet over medium heat for about 3 minutes, or until fragrant, stirring constantly. Remove the walnuts from the pan and set aside.

Now you're ready to assemble the salad. Put the farro in a large salad bowl and toss with 1 tablespoon of the dressing, the remaining ¼ teaspoon of salt, and pepper to taste. Next, add the beets, radishes, peas, and pea shoots and toss with the remaining dressing. Sprinkle with the walnuts and serve. Add a piece of burrata to each serving and more black pepper, if desired.

Vegan Zucchini Roll-ups

When zucchini are at their most plentiful in summer—and my neighbors keep dropping them at my door—they call for creativity. After all, there's only so much sautéed zucchini and zucchini bread one can eat. Enter these savory little vegan rolls: a bit of tangy vegan cheese with a hint of basil spread on grilled zucchini, rolled up and topped with tomato sauce. The dainty little rolls are surprisingly filling. Eat them alone or pair them with a fresh arugula salad for a refreshing lunch.

Stir the basil into the cashew spread and set aside.

Cut the zucchini lengthwise into very thin slices, about ¼ inch each. A mandoline works well here if you have one. Lay the slices out flat and brush each one on both sides with olive oil. Season each slice with a sprinkle of salt and ground pepper.

Preheat the oven to 350°F.

Meanwhile, heat a grill pan over medium heat. Grill the zucchini in batches for about 2 minutes on each side. Once all the zucchini have been grilled, lay out the slices and spread each one with about 2 teaspoons of the basil spread. Roll up each slice and place seam side down in a small baking dish (about 7 x 10 inches). The rolls can be touching. Pour your choice of sauce over the rolls. Place the rolls in the preheated oven and bake for 10 minutes, or until heated through. Garnish with basil leaves and serve immediately.

Serves 4 (makes about 20 rolls) | 12 g protein

- 1 tablespoon chopped fresh basil, plus more leaves for garnish
- ¾ cup Creamy Cashew Spread (page 43), or store-bought spreadable vegan cheese alternative
- 2 large zucchini
- 1 tablespoon extra-virgin olive oil
- ¼ teaspoon kosher salt
- Freshly ground black pepper
- 1 cup Chickpea Veggie Sauce (page 45) or jarred tomato sauce

Cook's Note:

In addition to My 12 Most-Used Kitchen Tools list (page 22), another item I highly value is my Lodge cast-iron reversible grill/griddle pan. It's about 16 inches long and covers two stovetop burners. It's great for indoor grilling (and the griddle side is perfect for pancakes!). For a non-vegan option, replace the Creamy Cashew Spread with ricotta or mascarpone cheese.

Crunchy Mung Bean Sprout Salad

So extremely easy, this salad is the perfect combination of crunchy vegetables. Sprouting activates and multiplies mung beans' nutrients, especially the B vitamins. Although sprouting beans takes a few days, it's very little hands-on time, and it's so fun to watch the sprouts grow—it's like a science experiment. It's also very satisfying to eat the fruits (or sprouts) of your labor. This is another recipe from my friend Ankita; she developed it to be served with the Kale and Carrot Tofu Scramble (page 83) for breakfast, but I just love it for lunch.

Make the dressing: In a small bowl, whisk together the lemon juice, tamari, salt, and olive oil.

Make the salad: Combine the radishes, carrots, jicama, mung bean sprouts, and scallions in a salad bowl. Drizzle the vegetables with the dressing and toss. Sprinkle with the parsley, salt, and pepper and let the salad marinate. Refrigerate for 1 hour or more before serving.

Serves 6 | 2 g protein

For the dressing:
2 tablespoons fresh lemon juice
4 teaspoons tamari
½ teaspoon kosher salt
3 tablespoons extra-virgin olive oil

For the salad:
1 bunch red radishes (about 6), thinly sliced (about 1 cup)
2 large carrots, peeled and cut into 2-inch matchsticks (about 2 cups)
1 small jicama, peeled and cut into 2-inch matchsticks (about 2 cups)
2 cups packed mung bean sprouts (page 27)
4 scallions, sliced, including the green parts (about ¼ cup)
⅓ cup fresh flat-leaf parsley leaves, chopped
¼ teaspoon kosher salt, or more to taste
¼ teaspoon freshly ground black pepper

Vegetable Noodles with Hemp Seed Basil Pesto

Have you been inspiralized? To turn zucchini, carrots, and many other vegetables into ribbons of noodlelike pasta substitutes, look no further than the "spiralizer" kitchen tool. Vegetable noodles are gluten-free, high in fiber, and less processed than traditional pasta. To boost the protein of this squash and carrot noodle dish, I make a hemp seed pesto to toss with the colorful spaghetti-like strands. Just 3 tablespoons of this nutty seed have 10 grams of protein and 9 essential amino acids. Because a plateful of this pasta is a nutritional powerhouse, you're sure to feel satiated—so get crankin'.

Serves 4 | 24 g protein

3 medium yellow squash

3 medium zucchini

3 carrots (at least 2 inches in diameter), peeled

1 teaspoon Garlic-Infused Oil (page 39)

1 teaspoon kosher salt

1 recipe Hemp Seed Basil Pesto (page 44)

1 pint heirloom grape tomatoes, cut into quarters

Hemp seeds, several fresh basil leaves, and shaved Parmesan cheese, for garnish

Cook's Note:

Instead of adding garlic outright, I infuse the oil, which makes this dish low-FODMAP as well.

Using a spiralizer, turn each squash and zucchini into noodles. Do the same with the carrots, keeping them separate from the squash and zucchini. Cut any of the noodles that are extremely long.

Heat the Garlic-Infused Oil in a large sauté pan over medium-high heat. Add the carrots and sauté for about 1 minute. Add the squash and zucchini and sauté for another 30 seconds, until just slightly softened. Sprinkle with the salt and transfer the cooked noodles to a fine-mesh colander to drain.

Dry the noodles between paper towels, place them in a large bowl, and toss gently with the pesto and tomatoes.

Sprinkle the pasta with hemp seeds, basil leaves, and shaved Parmesan cheese. Serve immediately.

Warm Buckwheat Salad with Apples

Buckwheat groats have a very peculiar flavor. Not that it's a bad one—personally I like it. It's intense, earthy, and curiously sweet. While buckwheat seems like a grain and is treated like one, it is actually a seed that is harvested from a flowering plant related to rhubarb. And despite the word wheat *in its name, buckwheat contains no trace of it. In fact it's gluten free. The kernels, which resemble steel-cut oats, are high in B vitamins and provide complete protein, as they contain all the essential amino acids. Make this salad when apples are at their crispest in autumn. Choose a variety of apple that's tart, such as Braeburn or Empire, to balance the sweetish buckwheat.*

Serves 4 | 12 g protein

For the dressing:

½ shallot, finely minced

2 tablespoons cider vinegar

2 tablespoons extra-virgin olive oil

½ teaspoon pure maple syrup

½ teaspoon Dijon mustard

¼ teaspoon kosher salt

⅛ teaspoon freshly ground black pepper

1 ounce goat cheese, softened

For the salad:

1 cup whole buckwheat groats

2 cups water

¼ teaspoon kosher salt

¼ cup pecan halves

3 cups arugula leaves

2 cups radicchio, thinly sliced
    (about ½ large head)

½ small red onion, thinly sliced

1 crisp red apple, cored and thinly sliced

2 ounces goat cheese, softened

Make the dressing: Put all the ingredients in a lidded jar and shake well. Set aside while you make the salad.

Make the salad: To cook the buckwheat, place the groats in a small saucepan with the water and salt. Bring the mixture to a boil, lower the heat, cover, and simmer for 10 to 15 minutes. Remove from the heat and let stand for 5 minutes. Fluff the buckwheat with a fork before using. Next, toast the pecans. Put the nuts in a dry skillet and heat them over medium-high heat, shaking the pan constantly, for 3 minutes, or until they start to smell toasted. Watch them closely as they can burn easily.

Spread the cooked buckwheat, arugula, radicchio, and onion on a platter. Shake the dressing well again and pour it over the salad. Toss to coat. Decoratively place the apple slices on the salad with spoonfuls of goat cheese and the pecan halves.

Cook's Note:

If you've never cooked buckwheat groats, watch the cooking time carefully. Buckwheat turns mushy very quickly when overcooked. Check your groats at 10 minutes in to see how much more time they may need, if any. If your buckwheat grains are tender and there is still water in the pot, simply drain it off.

Creamy Mushroom Barley Soup

The Natural Gourmet Institute (NGI) in New York City is more than a nationally accredited culinary school and educational center for nutrition and well-being—it's a community. I have attended many classes there, all of them supportive of healthful eating. This simple mushroom barley soup recipe comes from NGI. I love that the barley is puréed before it is added to the soup, which makes it creamy and smooth—without dairy. A good method to keep in mind when you want to thicken a sauce or soup—add in a little grain puréed with its cooking liquid.

Combine the barley, dashi, kombu, dried mushrooms (or reconstituted ones) and 1 teaspoon of the salt in a 6-quart stockpot over medium-high heat. Bring the mixture to a boil, then lower the heat and simmer, covered, for 1 hour. Remove the kombu after 30 minutes.

Put the olive oil in a medium-size skillet over medium heat and add the onion, sautéing for 5 to 8 minutes, or until softened. Add the carrot, celery, ½ pound mushrooms, 1 teaspoon salt and some freshly ground pepper. Lower the heat and continue to cook over low heat for 15 minutes.

Meanwhile, transfer the cooked barley, shiitake mushrooms, and cooking liquid to a blender jar and purée. Pour the purée back into the stockpot and add the sautéed vegetables, the remaining ¼ pound sliced mushrooms, and ½ teaspoon more salt. Simmer, covered, for 30 minutes.

Season with the remaining salt to taste, ground pepper, and a couple teaspoons umeboshi vinegar to taste. Garnish with the parsley.

Cook's Note:

If you make homemade stock and have reconstituted mushrooms leftover, use those in place of the dried mushrooms in this recipe. Also, I chop up the stems from the baby bellas and sauté them with the carrots, celery and mushroom caps (or you could reserve them for making stock). Lastly, for an even heartier soup, remove some of the cooked barley before you puree the stock and add it back in to the soup right before serving.

Serves 4 | 8 g protein

½ cup hulled, pearl, or hull-less barley

7 cups homemade Vegetarian Shiitake Dashi (page 36) or store-bought mushroom stock

1 (2-inch piece) kombu

1 ounce dried shiitake mushrooms

2 teaspoons kosher salt, divided

1 tablespoon extra-virgin olive oil

1 medium onion, diced (about 1½ cups)

1 large carrot, finely diced (about ¾ cup)

1 celery stalk, finely diced (about ⅓ cup)

¾ pound baby bella mushrooms, stemmed and thinly sliced, divided

1 tablespoon unsalted butter

Freshly ground black pepper

¼ cup chopped fresh flat-leaf parsley leaves

Umeboshi vinegar, for garnish (you may substitute red wine vinegar here; while it doesn't have the same sour and salty taste as the plum vinegar, it will still brighten the dish)

TLT (Tempeh, Lettuce, and Tomato) Sandwich

A play on the ubiquitous BLT (bacon, lettuce, and tomato), this sandwich replaces the bacon with tempeh. Like tofu, tempeh is also made from soybeans but because it's made from fermented whole beans, it has a firmer texture and retains more of its nutritional value (it's higher in protein and fiber than tofu). Firm and chewy, tempeh has a savory, earthy flavor that is the perfect complement to the crisp lettuce and fresh tomato.

Make the marinade: Mix all the ingredients together in a resealable bag. Add the tempeh and marinate, refrigerated, for at least 4 or up to 24 hours.

Make the sandwiches: When you're ready to make the sandwiches, remove the tempeh from the refrigerator and pat both sides of each piece dry with a paper towel.

Start your sandwiches by spreading four slices of the bread with the Lemon Mayonnaise and adding a piece of lettuce and tomato.

Next, put the olive oil in a skillet over medium-high and when the oil is heated, gently slide the tempeh into the pan, frying it for 3 to 4 four minutes on each side. Cook it in batches, if necessary.

When the tempeh is golden with somewhat crispy edges, remove it from the pan and transfer it to the prepared pieces of bread, topping each sandwich with a second piece of bread. Serve.

Cook's Note:

The Lemon Mayonnaise is key to this sandwich. It can be made up to 3 days in advance so you can have it ready when it comes time to assemble the sandwiches.

Serves 4 | 31 g protein

For the marinade:
3 tablespoons balsamic vinegar
1 tablespoon pure maple syrup
2 tablespoons tamari
1 garlic clove, minced
2 tablespoons extra-virgin olive oil
1 teaspoon Dijon mustard

For the sandwiches:
2 (8-ounce) packages soy tempeh, cut into ¼-inch thick strips
8 thick slices whole-grain sandwich bread
4 tablespoons Lemon Mayonnaise (recipe follows) or Lemony Aioli (page 40)
A few butter lettuce leaves
1 large tomato, sliced
2 tablespoons extra-virgin olive oil

Lemon Mayonnaise
⅓ cup store-bought mayonnaise
1 tablespoon fresh lemon juice
1 teaspoon lemon zest
1 garlic clove, minced
Stir all the ingredients together and use for the TLT Sandwiches.

Brown Rice Avocado Collard Wraps

The name of these wraps is misleading because they are so much more *than just brown rice and avocado. Crunchy, blanched collards are filled with beans, rice, mushrooms, fresh tomatoes, and red pepper. These flavorful, fiber-filled leaves make a great gluten-free alternative to traditional flour wraps.*

Serves 4 | 16 g protein

9 large collard leaves

1 tablespoon extra-virgin olive oil

½ small onion, finely chopped

1 garlic clove, minced

1 (8-ounce) container baby bella mushrooms, sliced

Kernels cut from 2 ears sweet corn

½ teaspoon kosher salt, or to taste

¼ teaspoon freshly ground black pepper

2 cups cooked brown rice

2 cups cooked or canned adzuki beans, drained

1 red pepper, seeded and cut into strips

1 avocado, pitted and sliced

2 tablespoons sesame seeds

For the collard wrap salsa:

2 cups grape tomatoes, cut into eighths

3 tablespoons fresh cilantro, chopped

1 tablespoon red onion, minced

1 tablespoon fresh lime juice

1 tablespoon kosher salt

Cook's Note:

Feel free to substitute another type of bean, such as black or pinto, for the adzuki, which just happens to be one of my favorites.

Bring a large pot of water to a boil. Cut away the thick part of the stem below the collard leaves. Place eight of the collard leaves in the boiling water and cook for 2 to 3 minutes, or until tender. Remove the leaves from the water and drain in a colander. Transfer the leaves to a baking sheet lined with paper towels, trying not to tear them. Finely chop the remaining uncooked leaf and set aside.

Meanwhile, make the salsa: Stir together all the ingredients for the salsa in a small bowl and set aside.

Heat the olive oil in a skillet over medium-high heat. Add the onion and garlic and sauté for about 3 minutes. Add the mushrooms and sauté for another 5 minutes. Add the corn, chopped collard, salt, and pepper. Stir to combine, and cook for another minute or two. Add the cooked rice and beans and toss the mixture thoroughly. Set aside to cool slightly.

Assemble the wraps: Place a softened collard leaf on a flat surface and top with one-eighth of the rice filling. Arrange a slice or two of the red pepper and avocado on top of the filling and a spoonful or two of the salsa. Sprinkle with a generous ½ teaspoon of sesame seeds.

Fold the two long sides of the leaf toward the center, and starting with the larger base end, roll the leaf like a burrito. Repeat with the remaining leaves and filling. Cut the rolls in half and serve.

Roasted Eggplant and Zucchini Wraps

I remember my first wrap well. It was from a now-defunct coffee shop in SoHo that made them to order. They would grill the freshly made flatbread, stuffed with herb-roasted fresh veggies and a thick slab of smoked mozzarella, in a panini press until the bread was almost crisp and the cheese had just started to melt. There was something about the flavors melding inside the toasted wrap that was especially toothsome. I still remember them today, all these years later, because I encounter many veggie wraps that are nothing more than dry white tortillas encasing lettuce, tomato, and maybe a piece of red pepper. Bleh, tasteless. I've tried to capture the spirit of that noteworthy rolled sandwich here—I hope you find it memorable, too.

Put two rimmed baking sheets in the oven and turn it on to 400°F to preheat.

Meanwhile, in a small bowl, stir together the olive oil, rosemary, garlic, salt, and pepper. Brush both sides of the eggplant and zucchini slices with the seasoned oil.

Remove the baking sheets from the oven and transfer the slices to the pans. Bake for 7 to 10 minutes, or until softened.

When ready to assemble, whisk together the tahini and buttermilk in a bowl. Add the arugula and toss. Next, lay out your wraps and place ¼ of the mozzarella and a big handful of arugula on each wrap. Top each one with ¼ of the eggplant and zucchini. To roll the wrap, fold in the two sides of the wrap, pull the side closest to you over the ingredients, and gently but tightly roll it up. You should have a cylinder (just like a burrito). Repeat this with the remaining three wraps. Heat a panini press or grill pan and heat the wraps for about 5 minutes. If using a grill pan, place a weight or heavy pan on top of the wraps to press them down. Allow the wraps to sit for few minutes before cutting in half or else the juices will run out.

Serves 4 | 23 g protein

2 to 3 tablespoons extra-virgin olive oil

2 teaspoons fresh rosemary, chopped

1 garlic clove, peeled and minced

¼ teaspoon kosher salt

⅛ teaspoon freshly ground black pepper

1 large eggplant (about 1 ½ pounds), cut into ⅛-inch slices

1 medium zucchini (about ¾ pound), cut into ⅛-inch slices

1 teaspoon tahini, store-bought or homemade (page 41)

1 teaspoon buttermilk

2½ ounces arugula (about 4 cups)

4 (8-inch) whole-wheat lavash or sandwich wraps

6 ounces fresh smoked mozzarella cheese, sliced

Cook's Note:

If you have one, a mandoline works well here for cutting the veggies very thin.

Grilled Vegetable and Fresh Ricotta Sandwich

Sun-dried tomatoes are an unsung hero. One cup of these little red beauties has 8 grams of protein and 7 grams of fiber, and is packed with vitamin C. Here I've used this concentrated source of nutrients in a pesto that is spread on bread with ricotta and topped with grilled vegetables. Close your eyes and let the Mediterranean flavors warm you any time of year.

Serves 4 | 13 g protein

½ pound asparagus
½ head radicchio, thinly sliced
1 small sweet Vidalia onion, thinly sliced
2 tablespoons extra-virgin olive oil
1 tablespoon balsamic vinegar
⅛ teaspoon kosher salt, or more to taste
Freshly ground black pepper
8 thick slices multigrain bread
½ cup ricotta cheese, homemade
    (page 42) or store-bought
⅓ sun-dried tomato pesto, homemade
    (page 44) or store-bought

Cook's Note:

While making homemade cheese sounds extremely intimidating, it's really very easy (at least when it's ricotta cheese). If you opt for store-bought, look for a fresh-style ricotta, which is extra creamy and rich, as opposed to mass-market ricotta cheeses.

Preheat a grill pan. Brush the asparagus, radicchio, and onion with the olive oil and balsamic vinegar and season with salt and pepper to taste.

Place the vegetables on the preheated pan and grill for 2 to 3 minutes on each side, or until softened.

Spread a piece of bread generously with a tablespoon or two of fresh ricotta, then a teaspoon or two of sun-dried tomato pesto, and top with the grilled vegetables. Serve immediately.

Favorite Tofu Bánh Mì

This is one of my favorite sandwiches—ever. It's the perfect balance of spicy and crunchy and its flavors are cool yet nicely acidic. Bánh mì is traditionally the term for a Vietnamese baguette sandwich (the baguette was a result of French colonialism) with a variety of fillings, from pork belly to egg. The combination here of crispy tofu with zesty pickled vegetables and lime mayonnaise on a crispy baguette is otherworldly.

Serves 4 | 17 g protein

For the pickled vegetables:
¼ cup cane sugar
¾ cup white vinegar
1 teaspoon kosher salt
¼ cup water
2 carrots, peeled and julienned
½ large daikon, peeled and julienned
1 small cucumber, julienned
2 jalapeño peppers, cut into rings and seeded

For the tofu marinade:
1 tablespoon tamari
2 tablespoons Asian fish sauce
1 teaspoon sriracha
1 tablespoon neutral oil, such as grapeseed or canola
1 (15-ounce) package firm tofu, sliced into ¼-inch slices

First, make the pickled vegetables: Stir together the sugar, vinegar, salt, and water in a large bowl and add the vegetables, cover, and allow to sit at least 2 hours but preferably overnight, refrigerated.

Make the tofu marinade: At least 2 hours before you are ready to eat but no more than 12, make the marinade. Mix together the tamari, fish sauce, sriracha, and oil in a small bowl. Lay the slices of tofu in a 9 x 13 glass baking dish and cover with the marinade. Turn the slices over in the marinade, cover with plastic wrap, and marinate in the refrigerator for at least 2 but up to 12 hours.

Make the lime mayonnaise: Mix the garlic, lime zest, and lime juice together with the mayonnaise. (Note if you use homemade aioli, no need to add the garlic as it's already quite garlicky!) Set aside.

Assemble the sandwiches: Preheat the oven to 400°F.

Remove the tofu from the marinade and blot dry with a paper towel, then sandwich the tofu between dry pieces of paper towel and place the baking dish over it and press gently to get out some of the moisture.

Before you cook the tofu, put the baguettes in the oven to warm for 5 minutes. Meanwhile, heat the remaining 1 tablespoon oil in a skillet over medium-high heat. Gently slide in the tofu and sear it on each side until somewhat crispy, 4 to 5 minutes per side, or until nicely browned.

Remove the baguettes from the oven and slice lengthwise. Remove some of the interior bread, if desired (see cook's note).

Spread a tablespoon of lime mayonnaise on each of the four bottom sections of baguette. Place a nice handful of drained pickled vegetables on top of the mayonnaise. Top each with ¼ of the crispy tofu slices and then ¼ of the cilantro, mint, and basil leaves. Finish by placing the other half of the baguette on top of each sandwich and serve.

For the lime mayonnaise:
1 garlic clove, minced (optional)
2 teaspoons lime zest
1 tablespoon fresh lime juice
4 tablespoons store-bought mayonnaise or Lemony Aioli (page 40)

For assembly:
2 (12-inch) baguettes, halved crosswise
1 tablespoon neutral oil, such as grapeseed or canola
½ cup fresh cilantro leaves
½ cup fresh mint leaves
¼ cup fresh basil leaves

Cook's Note:

If you like a ratio of more filling to less bread, after you cut the baguette lengthwise, with your hands, pull out some of the inside of the baguette. You could always just pile on more filling, but then the sandwich can become unwieldy!

Mushroom and Truffle Oil Frittata

One summer, while I was pregnant with my elder daughter, my husband and I traveled through France's Dordogne Valley, where truffles are a specialty. They are at once earthy, musky, and pungent—there's really nothing quite like them. Someone told me while I was there that in the Dordogne they place a fresh black truffle in a basket with eggs and let it sit overnight. By morning, the scent of the truffle has permeated the eggs and they are used to make truffle omelets. Because truffles are so pricey, I've opted for truffle oil or truffle salt to lend this frittata its unique taste. I always tease that the truffles I ate that summer may have permeated my daughter because she is a die-hard truffle lover, too!

Heat the olive oil in a 10-inch ovenproof skillet over medium heat. Add the shallots and sauté, stirring frequently, for 8 minutes, or until softened. Add the mushrooms and 1 tablespoon butter and toss together with the shallots. Lower the heat slightly and continue to cook the mushrooms another 13 to 15 minutes. Sprinkle with ½ teaspoon of the salt and ¼ teaspoon of the ground pepper.

Meanwhile, beat the eggs in a large bowl for about 1½ minutes. Add the milk, thyme, and remaining salt and whisk for another minute. Preheat oven to 350°F.

Add the remaining tablespoon butter to the mushroom mixture, still cooking over medium heat, and heat until it is melted. Stir to combine and spread out the mixture evenly over the bottom of the pan, using the back of the spoon. Pour the egg mixture into the pan and cook for about 1 minute. Sprinkle the Parmesan evenly across the top of the eggs. After about 4 minutes, place the pan in the preheated oven and cook for 15 minutes, or until cooked through but not dry. Remove from the oven and drizzle the truffle oil over the frittata (or sprinkle with truffle salt, if using) before serving.

Serves 4 | 14 g protein

- 1 tablespoon extra-virgin olive oil
- ½ cup thinly sliced shallots
- 10 ounces baby bella mushrooms, sliced
- 2 tablespoons unsalted butter, divided
- 1 teaspoon kosher salt, divided
- ½ teaspoon freshly ground black pepper, divided
- 8 large eggs
- ⅓ cup whole milk
- 1 tablespoon fresh thyme leaves, chopped
- ½ ounce Parmesan cheese, finely grated (about ⅓ cup)
- 1 to 2 teaspoons truffle oil, or ¼ teaspoon truffle salt, or more to taste

Cook's Note:

There is much controversy regarding truffle oil. Few natural truffle oils—olive or grapeseed oil infused with real truffles—are available. Instead, many producers use a chemical compound that mimics truffle aroma to flavor the oil. Domestically, Oregon Truffle Oil (oregontruffleoil.com) sells the real deal: US-grown truffles flavor their oil. You can tell if an oil is natural by the ingredients list. You'll see "natural truffle flavoring" or "truffle aroma" listed when synthesized gases are used. Alternatively, try truffle salt, which is scented with pieces of truffles.

Savory Spring Crostata

I adore the rustic feel of a crostata. The dough is quite forgiving. Just be sure you use parchment paper when you roll it, to prevent sticking. The addition of quinoa flour here adds a nutty, coarse texture to the crust. You may also want to use this crust and goat cheese as a base for other vegetables, perhaps roast squash in autumn or fresh tomatoes in summer. Adjust the herbs accordingly.

Serves 4 | 17 g protein

For the crust:
¾ cup white whole-wheat flour
½ cup quinoa flour (page 37)
¼ cup finely grated Parmesan cheese
1 teaspoon kosher salt
½ cup (1 stick, 4 ounces) unsalted butter, chilled, cut into 1-inch pieces
1 large egg, lightly whisked

For the filling:
4 ounces goat cheese, softened
1 teaspoon lemon zest
1 teaspoon fresh lemon juice
¼ cup chopped fresh chives
2 teaspoons chopped fresh mint
½ teaspoon fresh thyme leaves
¼ teaspoon kosher salt, plus more for seasoning asparagus
⅛ teaspoon freshly ground black pepper
½ cup shelled fresh peas
½ bunch asparagus (about ½ pound), trimmed
1 teaspoon extra-virgin olive oil

Cook's Note:
You may use all whole-wheat flour if you don't have quinoa flour.

Make the crust: In the bowl of a food processor, pulse the flours, Parmesan, and salt together. Add the butter and pulse again several times until coarse crumbs develop. Then, add the egg and run for 15 to 20 seconds, or until a ball of dough forms. Pat the dough into a 5-inch round disk and cover it with plastic wrap. Put the dough in the freezer for 10 minutes.

Meanwhile, make the filling: Preheat the oven to 400°F. In a bowl, mix the goat cheese with the lemon zest and juice, chives, mint, thyme, salt, and pepper. Set aside.

Bring a small saucepan of water to a boil and add the peas. Cook for 2 to 3 minutes, or just until tender. Drain, coarsely mash, and set aside. In another bowl, toss the asparagus with the olive oil and a good sprinkling of kosher salt.

Place the chilled dough disk on a piece of parchment paper, and using a rolling pin, roll out the dough into a 12-inch round, dusting with wheat flour as needed to prevent sticking. Transfer the dough round, with its parchment paper, to a baking sheet.

Spread the goat cheese mixture carefully on the dough round. Gently mash the peas into the goat cheese and top with the asparagus. Fold the edges of the dough up around the asparagus, creating a 1-inch border. Bake the crostata for 35 to 40 minutes, or until the crust is golden and crispy.

Creamy Cashew Soba Noodles

Buckwheat noodles never leave me feeling weighed down the way a bowl of white flour pasta can. Traditionally soba noodles are made from buckwheat flour, but sometimes they are made from a blend of flours that includes whole wheat. Look for soba noodles that are made from 100 percent buckwheat flour for a gluten-free option, and for one that is higher in protein (buckwheat has an amino acid score of 100, which means it's a very high-quality protein).

Make the noodles: Cook the noodles according to the package instructions. After you drain them once, rinse with cool water and drain them well again. Set aside.

Make the dressing: While the noodles cook, put all the ingredients for the dressing in a food processor and run until smooth.

Place the cooked noodles, carrots, cabbage, and scallions in a large salad bowl and pour the dressing over the noodles and vegetables. Use two forks or tongs to gently toss. Sprinkle with the cilantro and mint and toss again. Top with the cashews just before serving.

Cook's Note:

Try adding edamame, cucumber, or chopped bok choy in addition to or in place of the other vegetables in this dish. The more the merrier. You may want to increase the dressing by one-half if you're adding a significant amount of vegetables.

Serves 4 | 22 g protein

For the noodles:
- 8 ounces soba noodles
- 2–3 large carrots, spiralized or cut into 2-inch matchsticks
- 1 cup red cabbage, thinly sliced
- 3 scallions, sliced, including the green parts (about a heaping ½ cup)
- ⅓ cup fresh cilantro leaves
- ⅓ cup fresh mint leaves
- ½ cup roasted cashews, chopped

For the dressing:
- 2 tablespoons cashew butter
- 2 scallions, sliced, including the green parts
- 1 tablespoon fresh ginger, grated
- 1 garlic clove
- 2 tablespoons fresh lime juice
- 2 tablespoons toasted sesame oil
- 1 tablespoon tamari
- 1 tablespoon rice vinegar
- 2 tablespoons water
- ¼ teaspoon kosher salt

Brown Rice Balls (*Onigiri*)

From Italian arancini, *those delectable cheesy fried rice balls, to Chinese sweet dim sum rice balls, I'm a sucker for any kind of rice formed into a tasty bite. Onigiri, with their subtle flavor, are my current favorite. Traditionally a Japanese bento box staple, these rice balls are made from short-grain or sushi rice. Unlike sushi, though, onigiri rice is not seasoned with vinegar. Instead, it is formed around or with pickled ingredients, such as umeboshi (pickled plums) as a preservative.*

Today you'll find onigiri made with all sorts of ingredients. Here I've combined brown rice with a variety of healthy fillings. You can make one kind, all three, or even more. I also wrap a piece of nori around each one before eating. They are also perfectly portable for school or work.

Makes 12 rice balls | 15 g protein in 6 rice balls

2 tablespoons finely chopped cabbage

1 teaspoon rice vinegar

3 cups cooked brown short-grain or brown sushi rice

1 carrot, peeled, diced, and steamed

1 tablespoon minced scallion

2 tablespoons unhulled sesame seeds

½ avocado, cut into ¼-inch dice

½ cup firm tofu, cut into ¼-inch dice

1 sheet nori, cut into 1-inch pieces, plus more for serving (optional)

Kosher salt for forming the balls

Cook's Note:

If you have leftover *onigiri*, store them in the refrigerator but allow them to come to room temperature before serving. You can also heat them in a pan with a little sesame oil the next day.

Combine the cabbage and vinegar in a small bowl and set aside to marinate for 10 minutes, drain, and use as directed. Divide the cooked rice among three bowls (about 1 cup each). Add the other ingredients to each of the three bowls as described below. Toss gently to combine, then use the mixture to form balls.

Have a bowl of water set aside and make sure your hands are moistened before you roll each ball. Sprinkle your hands with the water and rub them together, then sprinkle them with salt. Next, put about 2 tablespoons of the rice mixture into your hand and form into a ball. Eat at room temperature. Wrap each ball with a piece of nori before eating, if desired.

Scallion and Carrot Rice Balls

1 cup cooked brown rice, steamed carrot, minced scallion, and 1 tablespoon sesame seeds

Cabbage and Avocado Rice Balls

1 cup cooked rice, marinated cabbage, diced avocado

Nori and Tofu Rice Balls

1 cup cooked rice, diced tofu, 1 sheet nori cut into strips, 1 tablespoon sesame seeds

Dinner

After a long, tiring day, there's nothing more rewarding than sitting down to a home-cooked meal. With today's busy schedules though, few have time to spend at the stove—especially during the week. I like to allot a few hours on the weekends to cook up a stew, broth, or grains in advance. With Vegetable Shiitake Dashi (page 36) in the fridge, it's easy to add some soba noodles and sliced vegetables. Or keep a container of Three-Grain Porridge (page 47) on hand to use as a base for mozzarella and tomatoes. Once you have a few choice basics in your repertoire, you can build on them with seasonal produce. The Beet and Cranberry Bean Farrotto (page 145) could be paired with any number of vegetables and legumes, and Portobello Mushroom with Freekeh and Artichokes (page 150) could be your blank palette. Use these recipes as inspiration to create your new favorites.

Spicy Three-Bean Chili

Even meat lovers will find this smoky chili satisfying. The beans and chunky vegetables give a hearty texture to this meatless mix, while mustard seed, molasses, and beer meld to deepen the flavors of the base. This chili improves when it's given a day for the flavors to mingle. It also freezes well, so oftentimes I double the recipe and save half for a quick dinner.

Serves 4 | 20 g protein

3 tablespoons extra-virgin olive oil

1 large onion, diced (about 2 cups)

3 large carrots, peeled and cut into ¾-inch dice (about 1¾ cups)

2 celery stalks, cut into ½-inch dice (about ¾ cup)

1 green bell pepper, cut into ¾-inch dice

2 jalapeño peppers, seeded and finely diced

3 garlic cloves, peeled and minced

1 teaspoon ground coriander

1 teaspoon smoked paprika

1 teaspoon chipotle powder

1 tablespoon ground cumin

1 teaspoon dried oregano

1 teaspoon kosher salt

¼ teaspoon freshly ground black pepper

1 (28-ounce) can fire-roasted tomatoes

8 ounces beer

2 cups vegetable stock

2 tablespoons tomato paste

1 tablespoon mustard seeds

2 teaspoons hickory liquid smoke

1 tablespoon molasses

3 cups cooked or canned mixed beans, such as black beans, red beans, kidney beans, and chickpeas

Fresh cilantro leaves, sour cream, and grated Cheddar cheese, for garnish

Heat the olive oil in a large Dutch oven over medium-high heat. Add the onion, carrots, and celery and sauté, stirring frequently, for 15 minutes, or until the carrots are softened (but not mushy) and charred on the edges. Next add the green pepper, jalapeño, garlic, and spices including coriander, smoked paprika, chipotle powder, cumin, oregano, salt, and pepper and stir well until all the vegetables are coated. Lastly add the tomatoes and their juices, beer, vegetable stock, and tomato paste, raise the heat, and bring to a boil stirring a few times to prevent sticking. Lower the heat and continue to simmer slowly.

Meanwhile, toast the mustard seeds in a dry skillet over medium-high heat, shaking the pan while they toast, and stop when they start popping, after about 3 minutes. They will darken slightly. Remove the seeds and grind them in spice grinder or with a mortar and pestle.

Add the ground mustard seeds, liquid smoke, molasses, and beans to the chili, bring back to a boil, and simmer for 10 minutes. Serve immediately, garnished with cilantro, sour cream, and Cheddar.

Skillet Cornbread

Bake up a batch of this rustic cornbread to serve alongside Spicy Three-Bean Chili (page 120).

Serves 6 | 7g protein

6 tablespoons unsalted butter, divided
1¼ cups white whole-wheat flour
¾ cup finely ground cornmeal
1 teaspoon baking powder
1 teaspoon cane sugar
½ teaspoon kosher salt
¼ cup plain yogurt
½ cup whole milk
2 large eggs
¼ cup fresh corn kernels

Put 2 tablespoons of the butter in an 8-inch round ovenproof skillet or baking pan. Place in the oven for the butter to melt and preheat the oven to 375°F while you prepare the batter. Check on the pan after 4 minutes and remove from the oven when the butter is melted.

Whisk together the flour, cornmeal, baking powder, sugar, and salt in a bowl and make a well in the center. In a small bowl, whisk together the yogurt, milk, and eggs until combined. Pour the wet ingredients into the well of the dry and stir until combined. Last, melt the remaining 4 tablespoons of butter and stir it and the corn into the batter until moistened. Remove the skillet from the oven and brush the melted butter around the skillet using a pastry brush. Pour the batter into the skillet, spreading it evenly in the pan, and bake for 20 minutes, or until golden and a toothpick inserted into the center of the cornbread comes out clean.

Roasted Kale, Green Bean, and Lentil Bowl

I never tire of eating at Candle 79, a comfortable, elegant vegan restaurant that occupies two floors of a townhouse on Manhattan's Upper East Side. The hummus there, sprinkled with smoked paprika, is one of my all-time favorites, as is the restaurant's famous Grilled Kale Salad. What follows is my re-creation of that dish. I'm not the only one who's tried copying it. I think anyone with a bent toward cooking who's tasted it may have tried to replicate it. The combination of vegetables, grains, and lentils is brilliant. This recipe is my homage to Candle 79, a place of inspiration (and joy!).

Preheat the oven to 375°F.

Using a strainer, rinse the lentils and remove any stones or foreign matter. Place the lentils in a saucepan with the bay leaf and cover with 3 to 4 inches of water (about 2½ cups but will depend on the size of your pan). Bring the water to a boil, then lower the heat and simmer, partially covered, for 20 to 25 minutes, or until the lentils are tender but still firm. (Keep an eye on the lentils to make sure they don't boil dry. Add more water, if needed.) Drain the lentils and set aside.

Next, place the turnips and greens beans on a large baking sheet and toss with 1 tablespoon of the olive oil and a generous sprinkle of kosher salt. Arrange in a single layer on the pan and bake for 20 minutes, or until tender, tossing once if necessary to prevent over browning. Remove from the oven when done and set aside.

While the turnips and green beans are roasting, put the kale on another baking sheet and toss with 1 to 2 teaspoons of the oil and a generous sprinkle of salt. Roast the kale as well for 8 to 10 minutes.

Serves 6 | 15 g protein

1 cup dried French green lentils

1 bay leaf

1 pound turnips, peeled and cut into ¾-inch cubes (about 3 cups)

¾ pound green beans, trimmed and cut into 1 ½-inch pieces (about 3 cups)

3 tablespoons extra-virgin olive oil, divided

Kosher salt

1 large bunch kale (about 1 ½ pounds), stemmed, cut in 2-inch strips

1 small red onion, thinly sliced (about 1 cup)

1 cup cooked grains, such as farro, barley, or wheat berries (optional)

Chive Vinaigrette (recipe follows)

Freshly ground black pepper

1 avocado, peeled, pitted, and cut into ½-inch cubes

½ cup roasted and salted sunflower seed kernels

CONTINUED

Remove the kale from the oven and put in a large salad bowl. Add the turnips, green beans, lentils, onion, and your desired grain, if using. Toss with several generous tablespoons of the Chive Vinaigrette (recipe follows) and sprinkle liberally with salt and pepper.

Test the salad to see whether you need more dressing, and add it, if desired. You may have some dressing left over if you're not adding a grain.

Finally, top with the cubed avocado and sunflower seeds.

Chive Vinaigrette: Place all the ingredients, except the olive oil and avocado, in a blender jar or food processor and process until blended. Slowly add the oil until the vinaigrette is emulsified. Add the avocado and process until blended. If needed, add more water, teaspoon by teaspoon, until your desired consistency is achieved.

Chive Vinaigrette
¼ cup minced fresh chives
¼ cup white wine vinegar
1 tablespoon Dijon mustard
1 tablespoon water
2 teaspoons honey
Kosher salt and freshly ground black pepper
¼ cup extra-virgin olive oil
1 avocado, peeled, pitted, and coarsely chopped

Stuffed Poblano Peppers

Roasting deepens the flavors of these longish, deep emerald chiles filled with a mélange of rice, beans, roasted sweet potato cubes, and a little bit of feta. Don't expect a cheese-laden Mexican-style poblano. These are not in that category. While the inspiration harkens back to those fried peppers, I've worked to lighten them up, highlighting the individual flavors of the filling. You can use most any grain-bean-veg combo here for the filling, such as quinoa–black bean–corn or millet–kidney bean–red peppers. These are great for a dinner party. Prepare them the day before and pop them in the oven when your guests arrive.

Serves 6 | 10 g protein

6 large poblano peppers

1 large sweet potato (about ¾ pound), peeled and cut into ¼-inch dice (about 3 cups)

1 tablespoon extra-virgin olive oil

¼ teaspoon kosher salt, divided

1 cup cooked short-grain brown rice

1 cup cooked or canned pink beans, drained

½ (15-ounce) can fire-roasted tomatoes, drained (a generous ¾ cup)

½ cup red onion, chopped

1 garlic clove, peeled and minced

¼ cup fresh cilantro leaves, chopped

½ teaspoon ground cumin

Freshly ground black pepper

¼ cup crumbled feta cheese (optional)

Cook's Note:

If you've never cooked with poblano peppers, don't be tempted to skip the step of removing their skins, which are waxy and tough and hard to digest. Also, you may substitute any bean here, including adzuki, pinto, or black.

Preheat the oven to broil. Lay the peppers on a foil-lined baking sheet and place directly under the flame for 5 to 7 minutes, or until the skins begin to char and blister, then turn them over with tongs and broil for another 5 minutes or so, until the other side begins to char. Transfer the peppers to a large bowl and cover the bowl tightly with a plate or plastic wrap.

Turn down the oven temperature to 375°F. Place the diced sweet potato on a parchment-lined baking sheet and toss with the olive oil and ⅛ teaspoon of the salt. Roast the sweet potato for 20 to 25 minutes, or until softened, tossing while cooking as needed. Remove the pan from the oven when the sweet potatoes are cooked through. Set aside. Lower the oven temperature to 350°F.

Meanwhile, after the peppers have cooled for 10 minutes, gently peel the loosened skin from them. Make a slit in the peppers and remove the seeds, being careful not to tear the peppers. Set aside.

CONTINUED

Put the cooked rice, beans, tomatoes, onion, garlic, cilantro, cumin, and remaining ⅛ teaspoon of salt in a bowl and mix until combined. Add a few turns of ground pepper, too. Next, add the roasted sweet potato and toss gently until incorporated.

Use ¼ to ½ cup of this mixture to stuff each pepper. Gently open the slit and place the mixture inside the pepper, making sure it fills the pepper, adding more if needed.

Place the peppers on a foil-lined baking sheet, sprinkle with the feta cheese, and bake at 350°F for 15 minutes. Serve immediately.

Sea Vegetable Brown Rice Bowl

Sea vegetable is a general term for a variety of edible seaweed, including nori (found wrapped around sushi rolls), kombu (used in broths), arame (thin, noodlelike texture), and dulse (with a high iodine and potassium content, it's a good salt replacement). Most sea vegetables are sold dried. While some, such as nori and dulse, are eaten that way, most are rehydrated. Using arame and dulse here impart a briny flavor to this fiber-rich stir-fry, which always leaves me feeling supercharged.

Put the arame in a bowl and cover it with cold water. Soak for 15 minutes, then drain and rinse. Set aside.

Whisk together 1 tablespoon of the toasted sesame oil with the hot pepper sesame oil, tamari, rice vinegar, miso paste, and water. Stir in the garlic and ginger. (You can also put all these ingredients in a food processer and run on high speed until smooth.)

Put the remaining tablespoon of toasted sesame oil in a large skillet or wok over high heat. Add the broccoli and carrots and sauté for about 2 minutes, stirring constantly. Add the cabbage and reconstituted arame and stir-fry for another 2 minutes. Pour 1 tablespoon of the sauce over the vegetables and toss to coat. Add the rice and stir well. Toss with the remaining sauce. Continue to cook for another minute or two until heated through. Sprinkle with the dulse flakes and sesame seeds before serving.

Serves 4 | 7 g protein

1 cup dried arame

2 tablespoons toasted sesame oil, divided

½ teaspoon hot pepper sesame oil

2 tablespoons tamari

1 tablespoon rice vinegar

1 tablespoon miso paste

3 tablespoons water

1 garlic clove, peeled and minced

¾ teaspoon fresh ginger, grated

2 cups broccoli florets

2 carrots, peeled and cut into 3-inch matchsticks

1 cup napa cabbage, thinly shredded

2 cups cooked short-grain brown rice

4 teaspoons dulse flakes

2 tablespoons toasted sesame seeds (page 41)

Cook's Note:

Unfortunately there really aren't any non-sea vegetable substitutions for arame and dulse. You can make this stir-fry without them, or if you can't find them in your local market, you can order them from an online retailer such as Maine Coast Sea Vegetables or Vitacost.

Soba Noodles in Broth with Bok Choy

Comforting like a familiar sweater, buckwheat soba noodles in broth are warming and soothing. I especially like a bowl of these on a cold night. Because the soup's light, it's both energizing and mood-elevating regardless of the time of the year though. The purity of the broth paired with an array of vegetables, including immune-boosting mushrooms, creates a harmonious blend. It's all very Zen.

Put the dashi in a large pot and bring to a simmer. Add the tamari, mirin, bok choy, and mushrooms and continue to simmer over low heat for about 5 minutes, or until the bok choy is just tender. Turn off the heat and cover to keep warm.

Meanwhile, heat another pot of water to cook the buckwheat noodles. When it comes to a boil, cook the noodles according to the package directions and drain when al dente.

Divide the heated stock, bok choy, and mushrooms among four large soup bowls. Then, divide the rest of the ingredients, including the sprouts, radish, scallions, and noodles, among the bowls, too. Serve piping hot.

Serves 4 | 10 g protein

8 cups Vegetarian Shiitake Dashi (page 36) or store-bought mushroom stock

4 tablespoons tamari

3 tablespoons mirin

2 heads baby bok choy, cut in half lengthwise with ends trimmed

6 to 8 shiitake mushrooms, thinly sliced

6 ounces dried buckwheat noodles (see headnote on page 115 regarding buckwheat noodles)

1 cup bean sprouts

4 radishes, thinly sliced

2 scallions, thinly sliced, including white parts

Cook's Note:

I discovered *Hon-mirin*, a sweet rice wine, several years ago and now use it almost exclusively in place of more commonly found *aji-mirin*, which is usually sweetened with corn syrup or glucose and often contains preservatives. *Hon-mirin*, sometimes referred to as "real" mirin, has a higher alcohol content and gets its sweetness naturally from fermented rice. Eden brand also makes a natural mirin without added sugar or synthetic enzymes.

Red Kidney Bean Stew

Spicy, rich, and flavorful, this Indian-style stew uses kidney beans, which are extremely high in fiber and protein, and a fabulous source of folate and iron, too. This is the perfect make-ahead meal because the flavors only get better as they intermingle for a night or two. Introduced to me by my friend Ankita, this dish is a staple in northern Indian homes. While working on this recipe, Ankita consulted her mother by phone several times in India, where red kidney beans are called rajma, *to ensure she was making it "right." It certainly tastes right to me.*

If you've never used hing, *also called asafetida, a spice this recipe calls for, don't drive yourself crazy trying to find it. The dish works just fine without it, but if you can get your hands on some, possibly from an Indian market, definitely use it. Made from tree resin, hing is used for its antiflatulent and digestive-aid properties as well as its distinctive flavor that adds a certain umami to foods. Its aroma is quite strong, though it will mellow when you sauté it.*

Last, you can substitute two serrano peppers for the Indian green chile, if necessary.

Sort, wash, and soak the kidney beans overnight in a large bowl or pot (the water level should be 3 to 4 inches above the beans).

The next day, drain the beans and put them in a pot. Cover them with about an inch of water and bring them to a boil, then lower the heat, cover, and simmer until the beans are tender, which could be 1 hour or more. When just softened, stir in ½ teaspoon of the salt, stir, and drain, reserving the cooking liquid. Set aside both the beans and the liquid.

Put the ginger, garlic, and chile in the bowl of a food processor and run until a smooth paste forms. Set aside. Heat the oil in a large Dutch oven and add the cumin seeds, *hing*, and bay leaves and stir well. Add the onion and sauté for 8 minutes, or until softened. Next, add the garlic paste and continue to sauté until the mixture turns golden, another 8 to 10 minutes. Finally, add the

Serves 4 | 7 g protein

- 1½ cups dried red kidney beans
- 1½ teaspoons kosher salt, divided, plus more to taste
- 1 (2-inch) piece ginger, peeled
- 7 garlic cloves, peeled
- 1 Thai or Indian green chile, seeded
- 2 tablespoons extra-virgin olive oil or ghee (page 39)
- 1 teaspoon cumin seeds
- ¼ teaspoon *hing* (optional)
- 2 bay leaves
- 1 large red onion, finely chopped (about 1¾ cups)
- 1 (15-ounce) can whole peeled tomatoes
- 1 tablespoon ground coriander

CONTINUED

¼ teaspoon red Indian chili powder or
   hot paprika
¼ teaspoon ground turmeric
1½ teaspoons garam masala
¼ cup fresh cilantro leaves, chopped,
   and lime wedges, for garnish
Cooked rice, for serving

Cook's Notes:

You can substitute canned red beans for dried ones in this recipe, but I encourage you to use dried ones if possible. Because this dish is all about the beans, you want them to have the best taste and texture possible.

tomatoes and their juices, crushing the tomatoes with your hand as you add them to the pot. Bring the mixture to a simmer, then add the coriander, red chili powder, and turmeric. Continue to simmer over low heat for another 10 minutes or so.

Add the cooked beans to the pot and slowly add ¼ cup or more of the bean-cooking liquid until a creamy consistency is achieved. Stir in a teaspoon of the salt and the garam masala. Stir well. Simmer for 15 to 20 minutes over low heat, or until heated through. Adjust the seasonings as needed. Garnish with the cilantro leaves and a squeeze of lime. Serve with rice.

Saag Paneer

Madhur Jaffrey was my first inspiration in the kitchen for Indian cooking. Well, not quite. It was actually my husband, Jim. It was he who introduced me to Jaffrey (not in person, unfortunately, just via her cookbooks). Jim and I both had an appetite for Indian food before we met. A one-time resident of London, my husband savored the many Indian cuisine options the city offered. Simultaneously I traveled to India and got a taste of the country's aromatic seasonings. When Jim and I met, we shared a love for many things—Indian cooking was just one of them. Suffice it to say, this dish is a favorite. Serve this spinach dish with Red Kidney Bean Stew (page 133).

Put the garlic, chile, ginger, and water in a blender jar or food processer and purée until a paste is formed. Set aside.

Heat 2 tablespoons of the ghee in a large skillet over medium-high heat. Working in batches, gently fry the cubes of paneer until golden. Set aside.

In the same skillet, add another tablespoon of the ghee and continue to heat over medium-high heat. Add the fenugreek and cumin seeds and fry for 30 seconds, stirring constantly. Add the onion and cook for 2 minutes. Lower the heat to medium-low and continue to cook for another 8 to 10 minutes. If the onions become too dry, add a tablespoon or two of water.

When the onions begin to brown slightly, add the garlic paste you made earlier and cook for another few minutes. Add the garam masala, cumin, and coriander. Stir well.

Finally, add the spinach, salt, ground pepper, and milk. Stir well and heat through. Transfer the spinach mixture to a food processor and run until nearly puréed. Finally, heat the last tablespoon of ghee in the same skillet and return the puréed spinach. Add the paneer, stir, and heat through. Serve immediately.

Serves 4 | 15 g protein

- 4 garlic cloves, peeled
- 1 serrano chile, stemmed and seeded
- 1 (1-inch) piece fresh ginger, peeled
- ¼ cup water, plus more if needed
- 4 tablespoons ghee (page 39) or unsalted butter, divided
- 6 ounces paneer, homemade (page 43) or store-bought, cut into 1 ½-inch cubes
- ½ teaspoon fenugreek seeds
- ½ teaspoon black cumin seeds
- 1 small yellow onion, finely chopped
- 1 teaspoon garam masala
- ¾ teaspoon ground cumin
- ¾ teaspoon ground coriander
- 1 (16-ounce) bag frozen cut spinach, defrosted, or 1 pound fresh spinach leaves, chopped
- 1 teaspoon kosher salt, plus more to taste
- Freshly ground black pepper
- ½ cup whole milk

Crispy Tofu Cauliflower Coconut Curry

If you like coconut milk, this dish is for you. The combination of curry paste and coconut milk here is divine, as in I-could-drink-the-sauce-it's-so-good divine. I always serve this curry over rice to soak up some of the delicious soup-like sauce. Friends were staying with us years ago and offered to make dinner. I asked if they needed anything from the store, and after checking my pantry and fridge, they asked for one ingredient: curry paste. They whipped up a curry very similar to this one, using whatever vegetables I had, and this dish has stayed in my dinner rotation since. Although they made it with chicken, I like to use tofu instead—it maintains protein levels sans meat.

Line a plate with several stacked paper towels. Remove the tofu from the package and drain. Place the tofu on the prepared plate, put a paper towel on top of it, and place a heavy pot or skillet on top of that. Allow the tofu to press for 10 to 15 minutes. After that, put the pressed piece of tofu on a dry paper towel, blot it dry, and season with a pinch of two or salt on each side. Cut the tofu into ¾-inch dice.

Heat 1 tablespoon of the olive oil in a Dutch oven over medium-high heat. When the oil is hot but not smoking, put in the tofu (it should sizzle). Without moving the tofu, allow it to get crispy and browned on one side, then flip it, using a spatula. Continue to fry until all sides are golden. Then, remove the tofu from the pan and set aside.

To the same pan, add the remaining tablespoon of olive oil, heat again, and add the onion. Lower the heat to medium and sauté for about 5 minutes, or until translucent. Add the potato, garlic, ginger, coriander, turmeric, cumin, and curry powder, and cook for 5 minutes, stirring frequently. Add the vegetable stock, bring to a simmer, and cook for about 8 minutes, or until the potatoes are softened. Next, add the green beans, red

Serves 6 | 11 g protein

- 1 (15-ounce) package firm tofu
- 2 tablespoons extra-virgin olive oil, divided
- 1 large yellow onion, finely diced (about 2 cups)
- 1 potato (about 8 ounces), peeled and cut into ½-inch dice (about 1½ cups)
- 4 garlic cloves, minced
- 1½ teaspoons ginger, grated
- 2 teaspoon ground coriander
- ½ teaspoon ground turmeric
- ¼ teaspoon ground cumin
- ½ teaspoon curry powder
- 2 cups vegetable stock
- 1 cup green beans, cut into 1½-inch pieces
- 1 red bell pepper, seeded and cut into 1-inch dice (about ½ cup)
- ½ head cauliflower, cut into florets (about 2 cups)
- 1 teaspoon kosher salt, plus more for seasoning
- 2 (13-ounce) cans coconut milk (about 3 cups)
- 2 tablespoons Asian fish sauce
- 5 to 6 tablespoons red curry paste
- Cooked rice, for serving
- Thai basil, for garnish

CONTINUED

Cook's Note:

If you can't get your hands on Thai basil, which can be found in some groceries now and many Asian markets, feel free to use sweet Italian basil. While Thai basil is slightly spicier and has a hint of anise, the Italian variety will still add a nice fresh note to the curry.

pepper, cauliflower, and salt and cook for 5 more minutes. Finally, add the coconut milk, fish sauce, curry paste, tofu, and remaining ½ teaspoon of salt and bring to a boil. Lower the heat and cook at a simmer for another 10 minutes or so, or until the cauliflower and green beans are cooked to your liking (I prefer the vegetables slightly al dente, so you may want to simmer the mixture a few minutes longer if you like yours more done). Serve over rice and sprinkle with julienned Thai basil.

Forbidden Black Rice with Pumpkin

I am a fan of black rice not only for its taste, but for its captivating dark purplish, nearly black color. This grain is a nutritional gold mine. Extremely high in the antioxidant anthocyanin (the same one found in blueberries), black rice is also a good source of iron, Vitamin E, and fiber. You may have also heard this grain called "forbidden rice," which it is called because in China it was once reserved only for nobles and Emperors. Adding yogurt to the water when cooking this dish gives it a very tangy taste, which is cut slightly by the sweetness of the golden raisins. This aromatic dish makes a simple yet hearty dinner when served with roasted pumpkin wedges and sautéed greens (page 31).

Get started by roasting the pumpkin (see the following page). Next, heat the ghee or coconut oil in a Dutch oven over medium heat. Add the onions and sauté for 5 to 7 minutes or until softened. Add the garlic and ginger and sauté another minute or so. Then add the rice, cardamom pods, turmeric, and salt and stir until everything is well combined and the rice is coated.

Next whisk together the yogurt and water until the yogurt is smooth and add it to the rice. Stir in the raisins and coconut as well and turn up the heat to bring the mixture to a simmer. When it begins to simmer, lower the heat, cover, and cook for 40 minutes. At that time, check the grains. They will always remain somewhat chewy, but you still want them to be tender. Add more water if needed. You may need to cook the rice as long as one hour. Allow the rice to sit covered for 10 minutes before serving, then fluff with a fork and sprinkle with coconut shreds and scallions.

Serves 4 | 11 g protein

- 2 tablespoons ghee (page 39) or coconut oil
- ¼ large yellow onion, chopped (about ½ cup)
- 2 garlic cloves, minced
- 2 teaspoons fresh ginger, grated
- 1½ cups black rice
- 3 whole cardamom pods
- 1 teaspoon ground turmeric
- 1 teaspoon kosher salt
- 1½ cup plain yogurt
- 1¾ cup water
- ½ cup golden raisins
- 2 tablespoons shredded coconut
- One scallion, sliced for garnish, optional

CONTINUED

To roast the pumpkin: Preheat the oven to 375°F and line a rimmed baking sheet with parchment paper. Then cut the pumpkin in half and remove the seeds, scraping the inside out with a metal spoon until it is clean. Next, cut the pumpkin into wedges and peel them.

Heat the oil in a large skillet over medium high heat. When the oil is hot, sear the pumpkin wedges in the pan for about 3 minutes per side. After that, place the wedges on the prepared baking sheet and season with salt and pepper, and drizzle with maple syrup.

Bake for 30 minutes or up to 50 minutes depending on the size of your wedges. When the wedges are fork tender, remove from the oven and spread the butter over the pumpkin so it melts right into the wedges. Serve.

For the pumpkin:
1½ pounds sugar pumpkin
1 tablespoon ghee or coconut oil
Kosher salt
Freshly ground pepper
1 tablespoon maple syrup
2 teaspoons butter

Roasted Acorn Squash with Quinoa

While I'm always sad to bid summer adieu, the prospect of cooler days and cozy nights is very appealing, as is the abundance of autumn produce that appears at our farm stand. When I spot the deep green rinds of acorn squash, I can't resist buying a few. One of my favorite preparations is filling half a squash with quinoa. Rich with vitamin C and fiber, the roasted squash is complemented by the whole protein of the light, nutty seed. A sprinkling of pomegranate adds a pop of color and crunch.

Combine the quinoa in a small saucepan with 1½ cups water. Bring to a boil. Cover, lower the heat, and simmer slowly for 15 minutes or until tender and most of the liquid is absorbed.

Preheat the oven to 400°F. Line an 11 x 17-inch baking sheet with foil. If needed, cut the bottom of each squash half so it sits flat. Next, place the squash halves face down on the sheet and add enough water to cover the bottom of the pan, about ⅛ inch, but not so high that it sloshes over the sides when you move the pan.

Place the pan in the middle rack of the preheated oven. After 30 minutes, using a spatula, gently flip the squash halves face up, one at a time, and rub the insides of each with butter, 1 teaspoon of maple syrup, and a few pinches of salt. Continue to roast for another 20 minutes, or until the squash halves are tender when pierced with a fork.

While the squash cooks, whisk together the lemon juice, vinegar, olive oil, salt, and pepper in a small bowl and set aside. In a larger bowl, gently stir together the cooked quinoa, pomegranate seeds, and scallions. Toss the mixture with the prepared dressing and divide among the roasted squash. Serve immediately.

Serves 4 | 20 g protein

- ¾ cup dried quinoa
- 2 acorn squashes, cut in half and cleaned
- ¼ teaspoon butter
- 4 teaspoons pure maple syrup
- 1 teaspoon kosher salt, plus more for seasoning
- 2 tablespoons fresh lemon juice
- 1 tablespoon cider vinegar
- 3 tablespoons extra-virgin olive oil
- ½ teaspoon freshly ground black pepper
- 1 cup pomegranate seeds (about 2 pomegranates)
- 3 scallions, finely sliced, including the white ends and about ⅓ of the green parts

Beet and Cranberry Bean Farrotto

Farrotto is a wonderfully creamy dish made in the same fashion as risotto (slow cooking with stock), but it replaces rice with farro. The whole-grain substitution results in a slightly chewier, more robust dish than the traditional rice one, while upping the nutritional ante.

The only unfortunate thing about this dish is that the stunning cranberry beans, also called borlotti or Roman beans, lose their brilliant color after cooking. Their sweet, nutty taste makes up for this loss, though.

Preheat the oven to 375°F. Put the farro in a bowl and cover it with boiling water. Let it sit for 20 minutes, then drain. Meanwhile, put the fresh beans in a medium-size pot and cover with cold water. Bring to a boil, lower the heat, and simmer for 20 to 25 minutes, or until tender. Drain and set aside.

Put the cubed beets on a large baking pan and toss with 1 tablespoon of the olive oil and sprinkle liberally with salt and pepper. Bake for 30 minutes, or until softened, tossing once or twice while roasting to prevent burning.

Next, heat the remaining 2 tablespoons oil in a Dutch oven over medium heat. Add the shallot and cook for 5 minutes, or until it softens, stirring frequently. Add the garlic and cook for another 30 seconds or so. Stir in the drained farro and mix well. Stir in ½ cup of the warm stock. Continue to stir until the farro has absorbed most of it. Then, add 3½ cups of the stock and bring to a gentle simmer. Lower the heat, cover, and cook for 40 to 50 minutes more, stirring occasionally and adding more stock as needed. There should be some liquid left with the grains, but not much. It should be creamy.

Stir in the cooked beans, beet greens, roasted beets, salt, and pepper and stir several times, until the farrotto is heated through. Last, sprinkle with the Parmesan and serve.

Serves 6 | 27 g protein

1½ cups uncooked farro

1¾ cups fresh cranberry beans (about 1½ pounds), shelled

¾ pound beets (1 small bunch), peeled and cut into ½-inch cubes, plus the beet greens, stemmed and chopped

3 tablespoons extra-virgin olive oil

1 teaspoon kosher salt, plus more for seasoning

Freshly ground black pepper

¼ cup shallot, diced (about 1 large shallot)

1 garlic clove, minced

4 cups vegetable stock, homemade (page 35) or store-bought, warmed until hot to the touch (140–150°F)

1 teaspoon fresh thyme leaves

½ cup coarsely grated Parmesan cheese (1½ ounces)

Cook's Note:

Cooking time is approximate here as it will depend on the brand and type of farro you use. Also, since fresh cranberry beans are only usually available for a short time, cook up some dried ones (pages 26–27) to use instead.

Roasted Brussels Sprouts with Maple-Sugared Orange Rind

Here, bright green Brussels sprouts are roasted in the oven until their outer leaves are slightly crispy and their insides are soft and tender. The baby cabbages become golden brown and glisten under a sweet orange glaze. A sprinkling of maple-sugared orange rind heightens the caramel flavor of the sprouts. If you don't want to take the extra few minutes to glaze the rind, you can substitute orange zest for it. These sprouts are dreamy paired with the White Bean Mash with Crispy Sage and Brown Butter (page 149).

Serves 4 | 4 g protein

1 pound Brussels sprouts, trimmed and halved lengthwise

1 shallot, thinly sliced

4 teaspoons extra-virgin olive oil

½ teaspoon kosher salt

Freshly ground black pepper

2 tablespoons pine nuts

1 tablespoon butter, unsalted

1 tablespoon white balsamic vinegar

2 teaspoons pure maple syrup

1 tablespoon fresh orange juice

1 teaspoon orange zest, or 3 tablespoons Maple-Sugared Orange Rind (recipe follows) (optional)

Cook's Note:

I give the option for the orange zest or candied orange rind because I don't want the orange flavor to overwhelm the others. If you love orange, though, go for it, use both.

Preheat the oven to 400°F. Toss the Brussels sprouts and shallot with the olive oil. Sprinkle with salt and pepper and transfer to a rimmed baking sheet. Roast the Brussels sprouts for 30 to 35 minutes, flat side down, tossing one or two times with a spatula. The sprouts should be nicely browned and the outer leaves crispy.

Meanwhile, toast the pine nuts in a dry skillet over medium heat, shaking it constantly, for about 2 minutes. Remove the nuts immediately from the pan to cool.

In the same pan, put the butter, vinegar, maple syrup, orange juice, and heat over medium heat, or until the mixture comes to a simmer. Lower the heat and cook for another minute or two, or until the mixture is slightly thickened.

When the Brussels sprouts are finished roasting, put them in a serving bowl, pour the warm butter mixture over them, and add the toasted pine nuts. Toss to coat.

Top with the orange zest or sugared rind and serve.

CONTINUED

Maple-Sugared Orange Rind

2 oranges

¼ cup pure maple syrup

3 tablespoons water, plus more for boiling

Maple-Sugared Orange Rind: Preheat the oven to 400°F. Line a small baking sheet with parchment paper.

Rinse the oranges and peel off the rind, using a vegetable peeler, trying to avoid the pith and only getting the orange skin. Next, cut the peels into thin strips, about ⅛ inch.

Fill a skillet with water and put in the orange rind. Bring it to a boil. Drain and repeat.

Put the drained orange rind, maple syrup, and the 3 tablespoons of water in the skillet and bring it to a simmer. Continue to cook over medium heat, stirring frequently, for 5 minutes, or until the mixture becomes thickened. Transfer the orange strips to the prepared baking sheet and spread out the mixture evenly. Allow it to cool for 15 minutes. Bake the peel for about 5 minutes. Remove it from the oven, allow it to cool, then break it apart as needed.

White Bean Mash with Crispy Sage and Brown Butter

When I crave something creamy and comforting, this bean mash satisfies that desire. While mashed potatoes and Brussels sprouts are a typical pairing, I like these beans even better. Plus, they pack a stronger protein punch. I think it's the drizzle of sweet browned butter and crunch of delicate fried sage leaves that make this mash seriously scrumptious.

In a skillet, melt 2 tablespoons of the butter over medium heat. Keep swirling the butter to make sure it melts evenly. The butter will start to bubble and foam, and after 3 to 4 minutes it will become a nutty brown color (and you'll see small brown bits separate out). At this time, remove the butter from the heat, pour it into a heatproof bowl, and set aside.

Melt the remaining tablespoon of butter in the same skillet and add the sage leaves. Stir-fry them for a minute or so, or until they become crispy. Remove them from the pan with a slotted spoon and set aside. Next, add the leek to the skillet and sauté, stirring occasionally, for about 5 minutes, or until softened. Add the garlic and sauté for 1 more minute.

Finally, add the beans, the reserved cooking liquid, salt, and pepper to taste and cook over medium heat, stirring frequently, until cooked through. At this time, you may either mash the beans in the pan with a fork or transfer the mixture to a food processor and purée for a smoother texture.

Transfer the mixture to a serving bowl and drizzle with the browned butter. Garnish with the crispy sage leaves and serve with Roasted Brussels Sprouts with Maple-Sugared Orange Rind (page 146).

Serves 4 | 10 g protein

3 tablespoons unsalted butter, divided

8 to 10 fresh sage leaves

1 large leek, thinly sliced (about ½ cup)

1 garlic clove, minced

3 cups cooked or canned cannellini beans

¼ cup bean cooking liquid, if needed

½ teaspoon kosher salt

Freshly ground black pepper

Roasted Brussels Sprouts with Maple-Sugared Orange Rind (page 146)

ortobello Mushrooms with Freekeh and Artichokes

Smoky-tasting, plump freekeh grains and creamy artichoke hearts make an aromatic hearty filling for portobellos, which are in turn a good source of protein, B vitamins, and other ailment-fighting compounds that make the mushroom unique. Meanwhile, freekeh boasts its own nutritional might. A longtime staple in Middle Eastern diets, this nutty, fiber-rich grain is wheat that's harvested when it's young and green, and is then toasted. You can buy it whole or cracked. It's a whole grain either way, but cracked freekeh takes less than half the time to cook. I prefer the whole grains for this recipe, but both work well.

Serves 4 | 20 g protein

- 5 to 6 large portobello mushroom caps, with rims if possible
- 1 tablespoon extra-virgin olive oil, plus more for brushing
- ½ teaspoon kosher salt, plus more for sprinkling
- 1½ cups cooked freekeh (or other cooked grain, such as wheat berries or farro)
- 1¼ cups or 1 (13- to 15-ounce) can artichoke hearts, drained and chopped
- ¼ cup finely grated Parmesan cheese
- 2 tablespoons fresh lemon juice
- ¼ teaspoon freshly ground black pepper, plus more for sprinkling
- 1 garlic clove, minced
- 1 teaspoon lemon zest
- 1 tablespoon fresh thyme leaves
- 3½ ounces fontina cheese, grated

Cook's Note:

You can assemble these mushroom caps one day ahead and keep them refrigerated until you're ready to bake them. I like serving this ample entrée with sautéed greens to guests.

Preheat the oven to 350°F and line a large, rimmed baking sheet with parchment paper.

Clean the mushrooms by brushing them gently with a damp cloth or soft brush. Snap off the stem, set it aside, and remove the gills by using the side of a spoon to gently scrape them off, lifting them away from the cap. Discard the gills.

Place the cleaned mushrooms on the prepared baking sheet. Brush them all over with olive oil and sprinkle with salt. Bake the mushrooms, top side down, for about 12 minutes, or until slightly softened.

Meanwhile, chop the reserved mushroom stems and place them in a bowl. Add the freekeh, artichoke hearts, and Parmesan. In a small bowl, whisk together the lemon juice, 1 tablespoon olive oil, ½ teaspoon of salt, and pepper. Stir in the garlic. Sprinkle the dressing over the freekeh mixture and toss. Sprinkle with the lemon zest and thyme leaves and toss again.

CONTINUED

After the first 12 minutes of baking, remove the mushrooms from the oven and divide the grain mixture among the mushrooms, spreading it evenly inside the caps. Return the mushrooms to the oven and continue to bake for 10 more minutes, then sprinkle with the fontina and bake for another 5 to 8 minutes (about 30 minutes total), or until the cheese is melted. Top with a sprinkle of ground pepper and serve.

Rich Lentil Stew

Inspired by misr wot, *a traditional Ethiopian lentil dish, this stew is rich and warming. However, this adaptation does not call for* berbere *(bear-BEAR-eh), a chili-based spice blend used in many Ethiopian dishes. Usually comprised of eight or more spices, including* ajwain, berbere *is rarely seen in the States. If you can find it, you can use it in place of some of the spices in this recipe (see cook's note). If not, don't worry—this dish is so pleasing without it. I use brown lentils because they hold their shape well (as opposed to authentic* misr wot, *which uses red). I like to serve this stew, which comes together quickly, over fluffy millet for a filling meal—plus a cup of it adds another 7 grams of protein to the meal.*

Using a strainer, rinse the lentils and remove any stones or foreign matter. Place the lentils in a saucepan with the bay leaf and cover with 3 to 4 inches of water (about 2½ cups but will depend on the size of your pan). Bring the water to a boil, then lower the heat and simmer, partially covered, for 20 to 25 minutes, or until the lentils are tender but still firm. (Keep an eye on the lentils to make sure they don't boil dry. Add more water if needed.) Drain the lentils, remove the bay leaf, and set aside.

Melt the butter in a Dutch oven over medium heat and add the paprika, coriander, cumin, and oregano. Sauté for a minute or so to season the butter. Add the ginger and garlic and stir well, cooking for another minute. Add the onion, tomato, and red pepper and cook for about 5 minutes, or until the onion is softened. Add the cooked lentils and the salt and cook over low heat until heated through, stirring frequently. Add the basil, stir well, cook for another minute, and serve over Fluffy Millet, seasoning with ground pepper to taste.

Serves 4 | 15 g protein

1 cup dried brown or green lentils

1 bay leaf

6 tablespoons unsalted butter or ghee (page 39)

3 tablespoons Hungarian paprika

1 teaspoon ground coriander

½ teaspoon ground cumin

⅛ teaspoon dried oregano

1 tablespoon fresh ginger, grated

1 tablespoon garlic, minced

1 medium red onion, finely chopped (about 1 cup)

½ cup tomato, diced

¼ cup roasted red pepper, chopped (page 28)

1 teaspoon kosher salt

¼ cup packed fresh basil leaves, julienned

Fluffy Millet (recipe follows)

Freshly ground black pepper

Cook's Note:

Three tablespoons *berbere* may be substituted for the paprika, coriander, and cumin in this recipe.

CONTINUED

Fluffy Millet: Put all the ingredients in a small pan over high heat. Bring to a boil, lower the heat to a slow simmer, and cover. Cook for about 15 minutes, or until the grains are tender. Try not to overcook as the grains will cook slightly more as they sit and steam. Let the pan stand covered, off the heat, for 10 minutes. After that time, fluff the millet with a fork and serve with the Rich Lentil Stew.

Serves 4 | 7 g protein

Fluffy Millet
1 cup uncooked millet
2 cups water or vegetable stock
½ teaspoon kosher salt
1 teaspoon unsalted butter

Crispy Three-Grain Cake with Mozzarella and Tomatoes

This clever recipe comes from my friend Rebecca Shim, who for many years specialized in vegetarian and vegan cooking as executive chef at Menla, a Tibet House retreat center. A square polenta-like cake made from a three-grain pilaf (amaranth, millet, and quinoa) is fried until it's crispy and golden. It's then covered with fresh mozzarella and topped with zesty tomatoes and red pepper. I also like to pair these cakes with some sautéed broccoli rabe or grilled eggplant and zucchini for a colorful, enticing plate.

Preheat the oven to 400°F.

Put the red pepper, tomatoes, and garlic in a bowl and toss with the vinegar, 1 tablespoon of the olive oil, and ⅛ teaspoon of the salt and pepper.

Heat the remaining 2 tablespoons of olive oil in a skillet over medium-high heat. Season the polenta squares with salt and gently slide them into the pan, cooking them for at least 5 minutes on each side, or until golden and crispy (add a little more oil if needed, to get a crispy crust). Transfer the cooked squares to a baking sheet and top each with a slice of mozzarella. Bake for 5 minutes or so, until slightly melted.

Put each square on a plate and top with the diced tomato mixture. Garnish with chopped basil and ground pepper to taste. Serve warm.

Serves 6 | 12 g protein

1 cup red bell pepper, seeded and diced

1 heaping cup grape tomatoes, quartered

1 garlic clove, minced

1 teaspoon balsamic vinegar

3 tablespoons extra-virgin olive oil, divided

1 teaspoon kosher salt, divided

⅛ teaspoon freshly ground black pepper, plus more to taste

6 (2½ x 3¼-inch) squares Three-Grain Porridge (page 47)

1 (8-ounce) piece fresh mozzarella, cut into 6 slices

Handful of fresh basil leaves, chopped, for garnish

Chickpea Ratatouille in Parchment

En papillote (on pa-pee-yoht), or "in parchment," is an easy technique that cooks food in a packet. Put a handful of raw vegetables, a cooked grain or legume (or both), and a sprinkle of seasoning in the center of a parchment paper heart and seal. The flavors of the ingredients meld together harmoniously as they steam inside. The sunny yellows, deep greens, and vivid reds of sweet summer vegetables make these particular little parchment parcels very cheery, but any seasonal vegetables will work. Add ½ cup of cooked grain to each packet if you'd like to up the protein.

Preheat the oven to 400°F. Cut four 15-inch square pieces of parchment and fold them in half. Cut out a large half-heart shape from each piece with the fold becoming the center of the heart. Set aside.

Put the zucchini and eggplant rounds, tomatoes, onion, and chickpeas in a large bowl. In a smaller bowl, stir together the olive oil, garlic, salt, and pepper to taste. Then sprinkle it over the vegetables with the herbs and hemp seeds and toss to coat.

Unfold the parchment hearts on two large, rimmed baking sheets. Place ¼ of the ratatouille mixture on ½ of a heart. Fold the other half over, lining up the edges of the heart and seal the parchment pack by starting at the top of the heart and making small creases around the edge of the heart as you work your way around to the tip of it. You'll be creating overlapping pleats. Fold one crease, move down and fold another crease, and continue. When you get to the bottom, twist the tip of the heart to seal the packet completely. Continue with the remaining parchment hearts and ratatouille mixture.

Place the packets on baking sheets (you'll need two large ones) and bake for 20 minutes. Serve the packets straight from the oven so they can be opened and enjoyed at the table.

Serves 4 | 8 g protein

- 1 small green zucchini (about 4 ounces), thinly sliced into rounds
- 1 small yellow zucchini (about 4 ounces), thinly sliced into rounds
- 1 small eggplant (about 4 ounces), thinly sliced into rounds
- ½ pint cherry tomatoes, cut in half
- ½ sweet yellow onion, thinly sliced
- 1½ cups cooked or canned chickpeas, drained
- 4 teaspoons extra-virgin olive oil
- 2 garlic cloves, minced
- ½ teaspoon kosher salt
- Freshly ground black pepper
- 1 teaspoon fresh thyme leaves
- 1 teaspoon fresh oregano leaves
- 1 teaspoon fresh rosemary leaves
- 2 tablespoons hemp seeds

Summer Tomatoes with Millet and Pesto Cream

It has taken me years, but I've at long last successfully grown tomatoes. More or less, anyway. A friend, Sarah, gave me a few starter cherry tomato plants one June that inspired me. I picked up a few more heirloom tomato plants at the farmers' market and filled half a bed with tomatoes. In the past I had always tried coaxing tomato plants to grow in containers but without much luck. Some little critter always got to them before they were fully ripened. So, that particular summer, into the ground they went. I was dedicated to my plants and they, in return, were fruitful. However, my tomatoes were still disappearing. I didn't understand. The garden is fenced. Then I spotted the culprit one afternoon when I looked out the window over the yard: Tess. Our Chesapeake Bay retriever is a stealth tomato thief. She ate every juicy ripe one and daintily left the green ones. Can you believe it? A dog who eats (and steals) tomatoes? I patched up the fence where she was sneaking in and managed to still harvest some vine-ripened beauties. Summer tomatoes are pretty irresistible, and need little enhancement to taste good, even to a dog, apparently.

When I have an excess of tomatoes because critters have been deterred, I like to get a little inventive and stuff them, like here, with millet. Paired with a salad of fresh garden greens, they're a meal that embodies summer.

Serves 4 | 19 g protein

½ cup uncooked millet

1½ teaspoons kosher salt, divided

1 cup water

4 large heirloom or beefsteak tomatoes, ripe but firm

½ cup fresh basil leaves, cut into julienne strips

½ cup spinach leaves, chopped

1 garlic clove, minced

¾ cup Parmesan cheese, divided

2 tablespoons hemp seeds

¼ teaspoon freshly ground black pepper

1 tablespoon extra-virgin olive oil

Pesto Cream Sauce (recipe follows)

Put the millet, ½ teaspoon of the salt, and the water in a small saucepan. Bring to a boil, lower the heat to a simmer, cover, and cook for 25 minutes, or until softened (there will still be a bit of a crunch). Set aside.

Meanwhile, preheat the oven to 375°F.

While the millet is cooking, prepare a baking sheet lined with paper towels. Set a wire rack on top of the paper towels. You will drain the tomatoes here. Next, hollow out tomatoes by cutting a large circle downward around the stem (almost to the rim of the tomato, as you would a pumpkin). Remove the top, cutting it loose, and then scoop out the pulp of the tomatoes into a bowl. Salt the

CONTINUED

insides of the hollowed out tomatoes and turn them upside down on the prepared baking sheet.

To make the filling, mash the tomato pulp you put in the bowl using a fork. Remove all but ½ cup of the mashed tomatoes and save them for another use. To the ½ cup of tomatoes, add the cooked millet, basil, spinach, garlic, ½ cup of the Parmesan, and the hemp seeds. Drizzle with olive oil and sprinkle with the remaining 1 teaspoon of salt and the pepper, or more to taste.

Line a baking sheet with parchment paper and set the tomatoes, base down, on it. Fill them with spoonfuls of the mixture, dividing it evenly among them. Top with the remaining ¼ cup of Parmesan. Bake for 25 minutes, or until warmed through, the tomatoes start to look juicy, and the cheese is melted.

While they bake, prepare the Pesto Cream Sauce. Serve the tomatoes straight from the oven with a few spoonfuls of sauce drizzled over each one.

Pesto Cream Sauce: Heat the butter in a small saucepan over medium heat. Add the half-and-half, then bring the mixture to a boil. Lower the heat to a simmer and whisk in the pesto, salt, and pepper. Continue to simmer for 5 minutes, or until the mixture is slightly thickened. Stir in the lemon zest, taste for seasoning (adding slightly more salt, pepper, or pesto, if needed), and serve warm over the Summer Tomatoes with Millet.

Pesto Cream Sauce
1 tablespoon unsalted butter
1 cup half-and-half
¼ cup store-bought pesto or Hemp Seed Basil Pesto (page 44)
½ teaspoon kosher salt
⅛ teaspoon freshly ground black pepper
1 tablespoon lemon zest

Jamaican-Style Collard Greens

When in Jamaica, my greens of choice are unquestioningly callaloo. The green, which is technically amaranth, is oftentimes compared to spinach, but I think it has a hardier texture than its sometimes wilty, more well-known cousin. Since callaloo is not easy to find near me, I've prepared collards in a similar way to how I've had callaloo prepared in Jamaica. One of these trips there I'm going to bring home some callaloo seeds to plant in my own backyard.

Heat the olive oil in a skillet over medium heat. Add the onion and cook for 5 to 7 minutes, or until softened. Add the garlic, vinegar, tomato, and whole Fresno pepper and cook for another minute or two. Finally, add the collards, coconut milk, and salt and bring to a simmer. Cover and cook over low heat for 8 to 10 minutes, or until the greens are wilted. Season with black pepper and serve.

Serves 4 | 2 g protein

1 tablespoon extra-virgin olive oil

½ large red onion, diced (about 1 cup)

2 garlic cloves, minced

¼ cup cider vinegar

1 tomato, chopped

1 red Fresno pepper

1 bunch collards, thick stems removed, thinly sliced

¼ cup canned coconut milk

¾ teaspoon kosher salt

Freshly ground black pepper to taste

Jamaican Rice and Peas

While it's not necessary to eat rice and beans at the same time to get the aminos you need for protein, you certainly can and should. They're a classic couple, like PB&J, cookies and milk, or even chips and dip. This combination of red kidney beans or pigeon peas with rice is a beloved dish of my father-in-law's, who is originally from Jamaica. His sister, Ouida, taught me to make them as they do on the island. Apparently, the contention over which bean to use is built into the dish. Every time we're in Jamaica, we find red kidney beans served with our rice. However, Ouida religiously makes them with pigeon peas—which my husband prefers over kidney beans. Our immediate family hasn't yet settled the debate over whether I should use pigeon peas or red kidney beans when I make it. Either way, the dish is still traditional—and delicious. And I've left the white rice here because, again, I've never eaten it any other way. One of these days I'll try brown rice, but if the competition between red kidney versus pigeon peas is any indication, that could open a whole new can of worms.

Heat the oil in Dutch oven over medium-high heat. Add the onion and cook for 5 minutes. Stir in the rice and beans and stir to coat with the onion mixture. Add the coconut milk, water, whole Scotch bonnet pepper, scallions, salt, ground pepper, and thyme. Bring the liquid to a boil, lower the heat to a simmer, and cover. Cook for 20 to 25 minutes, or until the grains are tender and the liquid is absorbed. Remove the scallions, thyme sprig, and Scotch bonnet. Serve immediately.

Serves 4 to 6 | 8 g protein (4 servings)

1 tablespoon extra-virgin olive oil

½ large yellow onion, diced (about 1 cup)

1½ cups uncooked long-grain white rice, such as basmati

1½ cups cooked or canned small red kidney beans or pigeon peas, drained

1½ cups coconut milk (about 1 [15-ounce] can)

1½ cups water

1 uncut Scotch bonnet or serrano pepper, whole

3 whole scallions

1 teaspoon kosher salt

½ teaspoon freshly ground black pepper

2 fresh thyme sprigs, or 2 teaspoons dried

Comforting Veggie Mug Pies

My daughter Camilla is passionate about potpies. One year she even asked to have them for her birthday dinner—which is in June. Obviously her desire for this savory pastry is not limited to the colder months.

I try to double this recipe so I can freeze half, which for Camilla, is better than money in the bank. The fennel and parsnip here give a hint of sweet to the creamy vegetable base, which gets a splash of color—and boost of protein—from the verdant edamame.

Serves 6 | 15 g protein

3 tablespoons unsalted butter

2 shallots, peeled and finely chopped (about ½ heaping cup)

1 leek, thinly sliced (about 1 cup)

2 russet potatoes, peeled and cut in ¾-inch dice (about 2 ½ cups)

2 large carrots, peeled and cut in ¾-inch dice (about 1 ¼ cups)

1 large parsnip, peeled and cut in ¾-inch dice (about ½ cup)

1 small fennel bulb, chopped (about 1 cup)

½ teaspoon kosher salt

Freshly ground black pepper

¾ cup vegetable stock, homemade (page 35) or store-bought

1 cup whole milk

¼ cup white whole-wheat flour

1 cup shelled frozen edamame

1½ teaspoons fresh thyme leaves

1 (1-pound) package puff pastry, defrosted if frozen, and cut into 6 5-inch squares

1 large egg, whisked with 1 teaspoon water

Cook's Note:

If you want to add some greens to this potpie, replace one of the potatoes with a cup or so of chopped curly kale. Also, consider adding mushrooms or celery.

Preheat the oven to 400°F.

Melt the butter in a Dutch oven over medium heat. Add the shallots and leek and sauté for 8 minutes, or until softened. Add the potatoes, carrots, parsnip, and fennel and continue to sauté for 5 more minutes, or until glistening. Sprinkle with the salt and freshly ground black pepper to taste, and toss to coat. Add the stock and bring to a boil. Lower the heat to a simmer.

Meanwhile, whisk together the milk and flour in a small cup until smooth. Add this mixture to the pot and raise the heat, bringing the mixture to a simmer again. Continue to heat, stirring, until the mixture is thickened. Stir in the edamame and thyme leaves.

Using a ladle, transfer the mixture to six 1-cup ovenproof mugs or mini casserole dishes, dividing it evenly among the mugs. Top each mug with a square of puff pastry and brush with the beaten egg mixture. Set the mugs on a foil-lined rimmed baking sheet and bake for 25 minutes, or until the pastry is puffed and golden. Let rest for 5 minutes before serving.

Savory Snacks

When that midafternoon hankering for a snack sets in, I like to have something nutritious on hand that will curb those cravings and keep my metabolism charged. Nut + Seed Protein Bars (page 179) or Roasted Edamame and Kale (page 181) always fill the bill. On weekends, when we have guests, I like to cook up a special snack a few hours before dinner to tide over hungry appetites. The small bites in this chapter are mostly that, something a little different from the everyday celery and hummus. They are meant to be healthy hors d'oeuvres or tasty morsels to be shared. Packed with fiber and protein, these flavorful nibbles will help keep energy levels—and spirits—high.

Red Lentil Hummus

Lentils are reported to increase longevity, which is reason enough to celebrate. This party-worthy dip is a good one to share with guests, but it also makes a wonderfully hearty snack. The fiber-rich legume helps keep your blood sugar levels even, which can be especially helpful in the afternoon if you're starting to feel fatigued from a busy day. Lentils may be small but they are nutritionally mighty and supply a great dose of folate, too. When they're paired with Seeded Crackers (page 172), you get 9 whopping grams of protein from this unassuming snack.

Put the lentils, stock, bay leaf, shallot, and garlic cloves in a small saucepan and bring to a boil. Lower the heat, cover, and simmer for 15 minutes. Remove from the heat and allow the lentils to stand, covered, for 5 minutes. After that, remove the bay leaf, drain off any excess liquid and transfer the lentils, shallot, and garlic to the bowl of a food processor. Add the remaining ingredients and pulse several times until the mixture is somewhat smooth but a slight texture remains. Transfer the hummus to a bowl and serve with Seeded Crackers (page 172).

Serves 8 | 3 g protein

1 cup dried red lentils

3 cups store-bought mushroom stock or Vegetarian Shiitake Dashi (page 36)

1 bay leaf

1 large shallot, peeled and sliced (about ½ cup)

2 garlic cloves, peeled

1 tablespoon red curry paste

¼ teaspoon ground cumin

¼ teaspoon ground coriander

1 tablespoon extra-virgin olive oil

2 teaspoons fresh lemon juice

1 teaspoon kosher salt

Seeded Crackers

Make these thin crackers to serve with Red Lentil Hummus (page 171), or most any dip. The variety of seeds adds B vitamins and tryptophan, an amino acid that makes serotonin, which helps promote calm and stabilize moods.

Makes 48 crackers | 6 g protein in 6 crackers

¾ cup buckwheat flour
¾ cup white whole-wheat flour
2 tablespoons poppy seeds
2 tablespoons raw sunflower seeds
2 tablespoons raw pumpkin seeds
¼ cup sesame seeds
1 teaspoon kosher salt
½ cup water, plus more if needed
2 tablespoons extra-virgin olive oil
1 tablespoon dark molasses
2 tablespoons caraway seeds
1 tablespoon coarse sea salt

Preheat the oven to 450°F.

In a medium-size bowl, stir together the flours, seeds, and kosher salt. Add the water, olive oil, and molasses and mix well until a sticky dough is formed. If the dough seems dry, add a tablespoon or two more of water until all the flour is moistened. Let the dough rest, covered, for 15 minutes.

Turn out the dough onto an 11 x 17-inch piece of parchment sprinkled with flour and form it into a rectangle. With a rolling pin, roll out the dough to be about ⅛-inch thick.

Lightly brush the dough with water and sprinkle with the caraway seeds. Press the seeds into the dough with the rolling pin. Next, using a knife or pizza cutter, cut the dough into 1 x 2-inch rectangles. Sprinkle with the sea salt and transfer the cut crackers along with their parchment to an 11 x 17-inch baking sheet.

Using a fork, prick the crackers all over. Bake for 15 minutes, or until the crackers start to brown on the edges. Remove from the oven and allow the crackers to cool completely. They will become even crispier as they cool.

Tamari Toasted Seeds

I always find a handful of these salty sauced seeds helps me stave off afternoon hunger. Problem is, it's hard to limit yourself to one handful. I even carry a small container of these when I'm on the go. They're great for a mid-trail snack or busy office afternoons—I like to pack them in my kids' lunch bags too.

Line a large rimmed baking sheet with parchment paper. Pour 2 teaspoons coconut oil in a large skillet and heat over medium high heat for a minute. Stir in the seeds and stir constantly for 3 to 4 minutes or until they start to smell toasted (don't wait for them to turn golden).

Remove the pan from the heat and toss the seeds with the tamari until coated (they will continue to cook in the hot pan so just watch that they don't burn). Spread the coated seeds out on the prepared baking sheet to cool and dry. Stir again after 30 minutes and eat or transfer to an air-tight container for storage.

Makes 2 cups | 8g protein in ¼ cup serving

2 teaspoons coconut oil
1 cup raw pumpkin seeds
1 cup raw sunflower seed kernels
2 tablespoons tamari

Roasted Paprika Chickpeas

Another addictive snack, these crunchy little bites are extremely satisfying. Many of you may have made roasted chickpeas in the past, but if not, once you start it's hard to stop. You can double or triple this recipe if you're making them for a large group. I keep my weekday batches small for two reasons: 1) I like them when they're fresh from the oven as they tend to get chewier as they cool, 2) I just keep eating them until they're gone! Like I said, these savory nibbles are addictive!

Make 1½ cups | 6 g protein in ½ cup serving

1½ cups cooked chickpeas
1 tablespoon extra-virgin olive oil
½ teaspoon kosher salt
1 teaspoon smoked paprika

Cook's Note

If you're using canned chickpeas, rinse them thoroughly under cool running water, discarding the liquid, and drain them well before blotting dry with a towel. You may substitute most any spice/herb for the smoked paprika.

Preheat the oven to 425°F.

Line a rimmed baking sheet with parchment paper. Lay out a few paper towels and blot the chickpeas dry discarding any loose skins. Put them on the prepared pan and drizzle with olive oil and, using your hands, toss to coat. Then sprinkle with salt and bake for 35 to 40 minutes or until crunchy.

When the chickpeas are finished roasting and you take them out of the oven, sprinkle them with paprika while still warm and toss to coat using a spatula.

Crunchy Sesame Seed Snacks

The modest sesame seed may be tiny, but it has nutritional might. The seeds are a stellar source of calcium and provide a good amount of protein too. Three tablespoons contain 5 grams of protein and about 24 percent of the USDA recommended daily calcium intake (a quarter cup of sesame seeds has more calcium than a cup of milk!).

You can usually find two types of seeds in stores: hulled and unhulled. Just as the name implies, the hull is removed from those seeds labeled "hulled." When the hull is removed, so is some of the calcium and iron—there is a slight loss of vitamins and minerals too. I almost always use unhulled because I like the texture, which is coarser than hulled. Like grains with their outer layers removed—say white rice—the end result is a smoother, more polished seed when you buy hulled. A great example: When making tahini (sesame seed paste), hulled seeds result in a smoother paste.

For this snack, where sesame seeds shine, you may use either. And don't get me wrong, hulled seeds still have their benefits. I just like to maximize my nutrition wherever I can.

Preheat the oven to 350°F.

Line a rimmed baking sheet with parchment paper. Put the egg whites in a bowl and whisk 1 to 2 minutes until nice and frothy. Add the sesame seeds, cheese, flour, salt, rosemary, and coconut oil. Stir together until thoroughly mixed. (Alternately, beat the egg whites in stand mixer fitted with a wire whisk for 30 seconds or until frothy, then add the ingredients as outlined above and run again for 30 seconds or until combined.)

Using a teaspoon, measure small rounds and place them on the baking sheet at least an inch apart. You should have 42 small rounds. Sprinkle with black pepper and poppy seeds.

Bake for 15 minutes or until golden. Remove from the oven and allow to cool before eating or storing in an air-tight container.

Makes 42 snacks | 7 g protein in 6 snacks

3 large egg whites
⅔ cup unhulled or hulled sesame seeds
½ cup coarsely grated Parmesan cheese
3 tablespoons white whole-wheat flour
½ teaspoon kosher salt
1 tablespoon dried rosemary
1 tablespoon aroma-free coconut oil
Freshly ground black pepper
1½ teaspoons poppy seeds

Spinach and Chickpea Spoon Fritters

When fried, these lovely little fritters become light on the inside and crispy on the outside. These pillowlike snacks are even better when dipped in Creamy Cilantro Avocado Sauce (recipe follows), but consider other dips as well, such as tzatziki or an herbed mayonnaise.

Put the flour, cornstarch, baking powder, cumin, coriander, salt, and ground pepper in the bowl of a food processor and pulse several times. Add the water and process until combined. Add the onion and Fresno pepper and pulse several times until combined. Next, add the chickpeas, spinach, and cilantro and pulse again until incorporated into the batter, but be careful not to overprocess. You don't want the leaves to become minced.

To fry the fritters, heat 2 to 3 inches of coconut oil in a deep skillet or Dutch oven over medium-high heat. Using a tablespoon, drop spoonfuls of the batter into the hot oil and fry for about 2 minutes on one side and one more on the other. When the fritters are golden, remove them from the oil with a slotted spoon and transfer to a paper towel–lined plate. Test one first to make sure it is cooked through. If not, lower the heat slightly and cook for another minute or two. Serve with Creamy Cilantro Avocado Sauce.

Serves 4 | 7 g protein

⅔ cup chickpea (garbanzo bean) flour

1 tablespoon cornstarch

½ teaspoon baking powder

½ teaspoon ground cumin

½ teaspoon ground coriander

½ teaspoon kosher salt

⅛ teaspoon freshly ground black pepper

½ cup water

¼ cup yellow onion, chopped

¼ red Fresno pepper, seeded and minced (reserve the remainder for the sauce)

½ cup cooked or canned chickpeas, drained

3 heaping cups fresh baby spinach

1 tablespoon fresh cilantro leaves

Aroma-free coconut oil, for frying

CONTINUED

Creamy Cilantro Avocado Sauce

1 cup fresh cilantro leaves

½ avocado, peeled

¼ cup fresh lime juice

¾ red Fresno pepper (left over from the Spinach and Chickpea Spoon Fritters)

2 tablespoons canned coconut milk

⅓ cup water

¼ teaspoons kosher salt

Freshly ground black pepper

Creamy Cilantro Avocado Sauce: Put all the ingredients in the food processor or a blender and process on high speed until smooth. Add more water or seasonings (salt and pepper), if necessary. Serve with the Spinach and Chickpea Spoon Fritters.

Nut + Seed Protein Bars

Homemade bars like these cost a fraction of what you'd pay for similar premade bars in stores. Equally advantageous is the fact you control all the ingredients for quality, sugar-content, and even preferred taste. Light and slightly chewy, these bars, with a balance of honey and cashews, can satisfy a sweet and salty craving. For a vegan option, replace the honey with pure maple syrup.

Preheat the oven to 350°F.

Cover the dates with boiling water and allow them to soak for 10 minutes. Meanwhile, grease an 8 x 8-inch baking pan with butter or coconut oil and line it with parchment so it overhangs on two opposite sides.

After the dates have soaked, drain them and put them in the bowl of a food processor and pulse several times until chopped. Add the cashews and coconut and pulse again. Transfer the mixture to a bowl and stir in the sesame, chia, and ground flaxseeds.

Put the honey, cashew butter, and salt in a small saucepan and heat over medium heat until it just comes to a simmer and is thinned and smooth. Remove from the heat and stir in the vanilla. Pour the warm sauce over the nut and seed mixture and stir until thoroughly combined.

Pour the mixture into the prepared pan and flatten with the back of a spatula. Bake for 25 minutes, or until it starts to turn golden. Remove from the oven and allow it to cool slightly. Use the overhanging edges to lift the parchment from the pan and transfer the bars to a cutting board. Cut into 16 bars.

Makes 16 bars | 5 g protein in 2 bars

½ cup pitted dates
Unsalted butter or coconut oil, for pan
½ cup unsalted roasted cashews
½ cup unsweetened shredded coconut
½ cup white and/or black sesame seeds
¼ cup chia seeds
2 tablespoons ground flaxseeds
¼ cup honey
2 tablespoons creamy cashew butter
¼ teaspoon kosher salt
½ teaspoon pure vanilla extract

Roasted Edamame and Kale

Even if I double or triple the recipe, this snack still doesn't make it off the baking tray. No matter who is in the kitchen when this kale-edamame combo comes out of the oven, he or she just can't seem to resist this toasty mixture. I rarely can even get it into a serving bowl. I wonder if this addictive snack, which I eat instead of chips, will disappear as quickly in your house. For reference, a single serving of potato chips contains about 2 ounces of protein.

Preheat the oven to 350°F.

Put the kale and edamame on a large, rimmed baking sheet and toss with the sesame oil, tamari and sesame seeds. Place in the oven and roast for about 8 minutes, or until the kale becomes crispy on the edges. Eat immediately.

Serves 4 | 7 g protein

1 bunch curly kale, stemmed, cut into bite-size pieces (3 to 4 cups)
1 cup shelled edamame
2 teaspoons sesame oil
2 teaspoons tamari
1 tablespoon sesame seeds

Carrot Mash Tartine

A few years back, a friend brought some savory squash toasts to our Thanksgiving feast. I swear I ate four before the meal started. I couldn't help myself. He'd made them from a piece Mark Bittman wrote in the New York Times—the recipe came from New York City's renowned ABC Kitchen. This carrot tartine is a riff on that original idea. The combination of carrot, coriander, and hazelnuts is divine but the real key here is a good cheese.

These toasts are meant to just be a little bite, an appetizer or afternoon snack to tide you over—although you can easily make a meal of them, as I almost did on Thanksgiving.

Serves 6 | 9 g protein

1 pound carrots (about 6 large carrots), peeled and cut into 1-inch rounds

1 tablespoon unsalted butter

2 teaspoons honey, plus more for drizzling

¼ teaspoon freshly ground coriander

¼ teaspoon kosher salt, plus more for seasoning

Freshly ground black pepper

6 to 8 pieces of whole-grain bread

1 teaspoon fresh lemon juice

1 teaspoon extra-virgin olive oil

1 cup pea shoots or other greens, such as arugula or watercress

½ cup fresh ricotta cheese (page 42)

3 tablespoons hazelnuts, chopped

Cook's Note:

If you can't find a small batch, homemade-style ricotta, you can make your own. It only takes two ingredients (milk and lemon juice) and very little effort, truly!

Place the carrots in a medium-size saucepan and bring to a boil. Lower the heat and simmer for 25 to 30 minutes, or until the rounds are fork-tender. Drain the carrots and return them to the saucepan with the butter, honey, coriander, salt, and pepper to taste (a few good grinds on a pepper mill will do). Mash coarsely with a potato masher or large fork, incorporating everything together.

Meanwhile, toast the pieces of bread.

In a small bowl, whisk together the lemon juice and olive oil. Toss with the pea shoots or other greens and sprinkle with salt and pepper.

To assemble the toasts, divide the ricotta among the pieces of toast (about a generous tablespoon on each) and spread evenly on each toast. Top with a thick layer of carrot mash and a small handful of greens. Sprinkle with hazelnuts and drizzle with a tiny bit of honey, if desired. Cut the slices in half. Serve immediately.

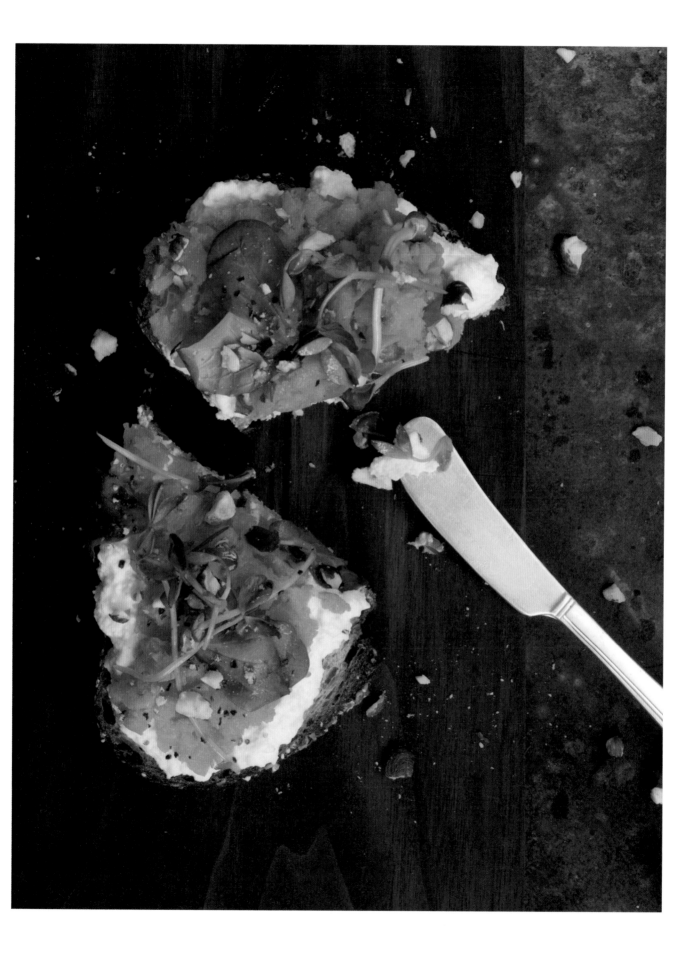

Scallion Sesame Pancakes

Our friend Julia, originally from China, stayed with us one weekend. Sunday morning, she taught us to make scallion pancakes her way. Unlike other more complicated recipes I knew that yield a dough, this one yields a batter. It comes together quickly, and Julia taught my daughter Camilla to whisk it with chopsticks, which Julia swears is the secret to a good pancake. This savory snack is made even more enjoyable with a dab of Kimchi Mayonnaise (recipe follows).

Using chopsticks, or a spoon, stir the flours and salt together in a bowl. Add the water, egg, tamari, sesame oil, sesame seeds, and scallions. Whisk together (again using chopsticks if you have them) until the mixture is just combined. Allow the batter to rest for 10 minutes.

Meanwhile, heat 2 tablespoons of the oil in a skillet or on a griddle over medium-high heat. Using a ¼-cup measure, pour the batter into the heated pan, making 5-inch rounds.

Cook the pancakes for 3 to 4 minutes on one side before flipping and cooking the other side for about the same amount of time, or until golden brown in spots.

Pour the remaining 2 tablespoons of oil into the pan after you have made half of the pancakes and wait until it is sufficiently heated (it should be hot but not smoking) before you begin with the next round of pancakes. Serve immediately with Kimchi Mayonnaise.

Kimchi Mayonnaise: Combine all the ingredients together in a small bowl and serve with Scallion Sesame Pancakes.

Makes 10 (5-inch) pancakes | 7 g protein in 2 pancakes

1⅛ cups white whole-wheat flour
⅛ cup arrowroot flour
1 teaspoon kosher salt
1 cup plus 6 tablespoons water
1 large egg, lightly whisked
1 tablespoon tamari
1 tablespoon sesame oil
1 tablespoon sesame seeds
4 scallions, including green parts, thinly sliced
¼ cup sesame oil for frying

Kimchi Mayonnaise
¾ cup mayonnaise
½ cup kimchi, chopped
2 teaspoons tamari
2 teaspoons rice vinegar

Sriracha Deviled Eggs

Perfect for a picnic, deviled eggs are an oldie but goody. A party staple, these spicy eggs also hit the spot any afternoon a protein boost is needed. No need to reserve these colorful eggs for guests; whip some up for a snack any day.

Serves 6 | 12 g protein

6 large eggs
¼ cup mayonnaise, store-bought or
  homemade (page 40)
1 teaspoon Dijon mustard
1 tablespoon sriracha sauce
¼ teaspoon fresh lime juice
¼ cup chopped roasted red pepper
Smoked paprika, for garnish

Cook's Note:

You may use either jarred roasted red pepper, or see page 28 for how to roast your own. If you use jarred peppers, be sure to drain them and pat them slightly dry with a paper towel.

Place the eggs in a single layer on the bottom of a large pot and fill it with enough water to cover them by 2 inches. Place the pot over high heat, bring the water to a boil, remove the pot from the heat, and cover. Allow the eggs to sit in the water for 13 minutes. Meanwhile, prepare an ice bath. After the eggs have rested in the water, drain them and place them in the icy water for 5 minutes. Drain them again and peel.

After the eggs are peeled, cut each one in half and gently scoop out the yolks into the bowl of a food processor. Add the mayonnaise, mustard, sriracha, lime juice, and red pepper. Pulse several times, until the mixture is combined but a slight texture remains. Scoop the filling back into the hard-boiled egg halves and arrange decoratively on a plate. Sprinkle with smoked paprika and serve.

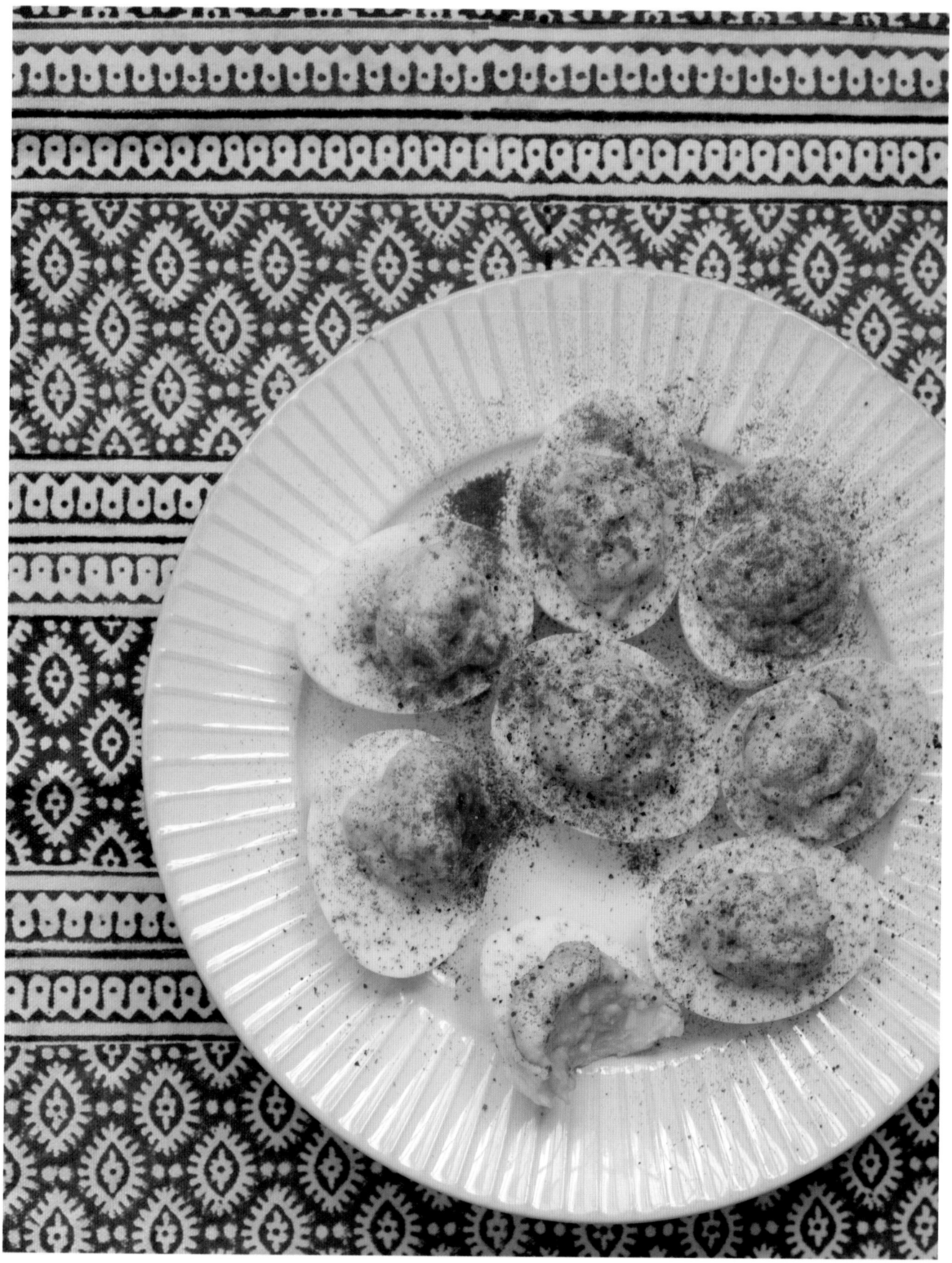

Sweets

Hopefully you find the sweets in this chapter intriguing and not off-putting. "Off-putting?" you ask. "Why would a sweet be that?" Because I know the thought of vegetables or whole grains in dessert does not sound appealing to everyone. I promise you, the addition of beets to a chocolate cake or brown rice to a berry crisp only enhance their flavor, making them richer, creamier, and best of all, more satisfying and nutritious. The desserts in this chapter are homey and comforting and incorporate wholesome ingredients, such as fragrant cinnamon, lush coconut milk, and golden maple syrup that make every sweet bite worthwhile.

Just think of veggies in dessert as "dirt candy," which I like to call them after Amanda Cohen's well-known vegetarian restaurant of the same name in New York City.

Maple Baked Apples

Special enough for a dinner party yet easy enough for a weeknight dessert, baked apples are the essence of autumn. Tart apples softened by the heat of the oven are complemented by a wholesome nutty filling. It just takes a minute or two to stir together the filling, but you can always substitute granola for an even speedier dessert. A scoop of ice cream, whipped coconut cream, or yogurt amplifies the comfort factor tenfold.

Preheat the oven to 350°F.

Using a knife, cut the bottom of each apple so it sits flat if it doesn't already. Then cut a small circle, about an inch or two in diameter, in the top of each apple and scoop out the core. Set the apples in a baking dish.

In a small bowl, stir together the oats, nuts, cinnamon, nutmeg, and salt. Then, stir in the maple syrup and vanilla. Last, stir in the butter.

Divide the filling among the six cored apples, packing it in tightly and mounding any excess on top.

Cover the dish loosely with aluminum foil and bake for 30 minutes. After that time, remove the foil and bake for another 15 minutes (about 45 minutes total), or until the apples are fork-tender. Serve warm with whipped cream, Coconut Cream (page 201), vanilla ice cream, or yogurt.

Makes 6 baked apples | 4g protein

6 Granny Smith apples
1 cup old-fashioned rolled oats
½ cup chopped nuts (walnuts, pecans, etc.)
½ teaspoon ground cinnamon
¼ teaspoon freshly grated nutmeg
Pinch of kosher salt
¼ cup pure maple syrup
½ teaspoon pure vanilla extract
4 tablespoons unsalted butter, at room temperature

Cook's Note:

Choose a good, firm baking apple, such as Granny Smith, when making this dish. If you use an apple like Red Delicious, it will most likely turn mushy.

Roasted Sweet Potato Pudding

The scent of cinnamon and nutmeg permeates the air while this fragrant pudding bakes, which is reminiscent of a pumpkin or sweet potato pie, just without the crust. The sweet potato becomes creamy and smooth when whipped with rich egg yolks and cream. This dessert transports well and is good for a crowd. You can also halve the recipe and bake individual servings in ramekins.

Serves 8 | 5g protein

4 to 5 medium sweet potatoes (about 2½ pounds)
6 large eggs
¾ cup heavy cream
⅓ cup pure maple syrup
1½ teaspoons pure vanilla extract
1½ teaspoons baking powder
2 tablespoons white whole-wheat flour
¾ teaspoon ground cinnamon
¾ teaspoon freshly grated nutmeg
¼ teaspoon ground cloves
¼ teaspoon kosher salt
More maple syrup and chopped toasted pecans, for serving

Cook's Note:

If you don't have a 7 x 10 pan, opt for something on the smaller side rather than the larger. I prefer the pudding thicker as it seems richer and more decadent.

Preheat the oven to 375°F. Grease a 7 x 10-inch rectangular or oval baking dish and set aside.

Line a rimmed baking sheet with foil. Rinse the sweet potatoes, prick them all over with a fork, and set them on the foil-lined sheet. Roast them for 45 minutes, or until fork-tender. Remove the potatoes from the oven and allow them to cool slightly.

When the potatoes have cooled enough to handle, scoop the softened flesh into the bowl of a food processor. Run until the potato is puréed. You should yield about 2½ cups. Add the eggs, cream, maple syrup, vanilla, baking powder, flour, cinnamon, nutmeg, and cloves and pulse again until smooth.

Pour the mixture into the prepared dish. Bake for 25 minutes still at 375°F, or until the mixture is golden. Serve spoonfuls in a bowl, drizzled with maple syrup and sprinkled with toasted pecans.

Dark Chocolate Beet Cakes

Rich and chocolaty, these gluten-free cakes are decadent enough for a special occasion. Puréed beets, which are barely detectable, add sweetness and a few extra grams of fiber and protein. Plate these cakes with a scoop of vanilla bean ice cream and you've got an elegant ending to even the most extravagant meal.

Preheat the oven to 350°F. Butter and flour (using almond flour) a 12-cup muffin tin or mini Bundt cake pan.

Boil the beets until soft, 35 to 40 minutes, then drain and purée, using a blender or food processor. Add 1 tablespoon cooking water to the beets if necessary. Set aside to cool.

Melt the butter and chocolate in a bowl placed over a small pan of simmering water. (Do not let the water get into the chocolate.) Remove the melted mixture from the heat and add the vanilla. Set aside to cool.

Meanwhile, in the bowl of a stand mixer fitted with the paddle attachment, beat the eggs and sugar together for about 10 minutes, or until thick and pale. When you remove the paddle, thick ribbons should form from the batter coming off the paddle. Fold in the almond flour and the salt.

Stir the puréed beets into the chocolate mixture and fold that mixture into the egg mixture until it is fully incorporated. The mixture should be consistently dark and have no striations. Finally, stir in the chocolate chips. Pour the batter into the prepared pans.

Bake for 20 to 25 minutes, or until a toothpick inserted into the center comes out clean. Remove the pan from the oven and allow to cool for about 10 minutes. At that time, turn over the pan to release the cakes. Serve warm with a dusting of confectioners' sugar.

Serves 6 | 11 g protein

½ cup (1 stick, 4 ounces) unsalted butter, plus more for pan

½ cup almond flour, plus more for pan

2 beets (about 8 ounces), peeled and quartered

8 ounces (½ pound) bittersweet chocolate

2 teaspoons pure vanilla extract

4 large eggs, at room temperature

¼ cup cane sugar

½ teaspoon kosher salt

¼ cup bittersweet chocolate chips

Confectioners' sugar, for dusting

Spiced Squash Snack Cake

Everything about a spiced snack cake evokes hominess: the pleasingly rich, honey brown color, the familiar soft texture, and the fragrant, spice-infused aroma—which to me is the scent of the holidays. This moist cake, made even tastier with the addition of butternut squash, has staying power, which makes it great for gift giving.

Preheat the oven to 350°F. Grease a 9-inch round springform pan. Line the bottom of the pan with a round of parchment.

In a small bowl, combine the flour, salt, cinnamon, ground ginger, cloves, nutmeg, and ground pepper. In another small bowl, dissolve the baking soda in the very hot water. Set aside.

In the bowl of a stand mixer fitted with the paddle attachment, combine the squash and oil. Add the maple syrup and molasses and mix thoroughly on medium speed. Add the eggs, continuing to mix until fully incorporated. Stir in the fresh ginger. Slowly add the flour mixture and beat on low speed until fully incorporated, about 40 seconds. Add the baking soda mixture. Pulse the mixer a few times on low speed so the water doesn't splash all over, then run at low speed until completely incorporated.

Spoon the batter into the prepared pan. Bake for 30 to 35 minutes, or until a toothpick inserted into the center of the cake comes out clean. Let cool completely in the pan. Gently run a knife between the edge of the cake and pan before opening. If desired, place confectioners' sugar in a sifter and sprinkle over the cake.

Serves 8 | 6 g protein

Unsalted butter or coconut oil, for greasing the pan

2¼ cups white whole-wheat flour

½ teaspoon kosher salt

1 teaspoon ground cinnamon

1½ teaspoons ground ginger

½ teaspoon ground cloves

¼ teaspoon freshly grated nutmeg

¼ teaspoon freshly ground black pepper

2 teaspoons baking soda

¼ cup very hot water, about 120°F

1 cup butternut squash, puréed (see Cook's Note)

½ cup coconut oil or unsalted butter, melted

¼ cup pure maple syrup

½ cup unsulfured dark molasses

2 large eggs

2 teaspoons fresh ginger, grated

Confectioners' sugar, for dusting (optional)

Cook's Note:

To make butternut squash purée, cut a squash in half lengthwise, remove the seeds, and roast at 400°F for up to one hour, or until fork tender. After roasting, scoop out the softened insides and purée in a food processor.

White Whole-Wheat Strawberry Shortcakes

Strawberry shortcakes scream "summer." The best time to eat this dessert is when berries are juicy and red. You can always use the shortcakes as a base for other fruits in winter, but it makes them extra-special when you know these darlings are seasonal. Just like summer days that come and go, I appreciate every bite of these shortcakes because I understand they, too, are fleeting.

Make the cakes: Preheat the oven to 425°F. Line a rimmed baking sheet with parchment paper.

In a medium-size bowl, stir together the flours, baking powder, sugar, and salt. Next, drop the chilled pieces of butter into the flour mixture and toss gently with your fingers. Using two forks or a pastry cutter, work the butter into the flour mixture until it resembles coarse crumbs. In a small bowl, whisk together the milk and egg. Pour into the flour mixture and stir well. Then use your hands to work the shaggy mass into a soft dough.

Turn out the dough onto a floured surface and press it down with floured fingers to about a ¾-inch thickness (approximately a 7 x 5-inch rectangle). Cut out rounds, using a biscuit cutter or glass (if you use a 2½-inch cutter you'll yield eight rounds). Brush the rounds with the melted butter. Bake for 14 minutes, or until golden and remove from the oven.

Meanwhile, make the topping: Slice the strawberries and put them in a bowl with the coconut sugar and lemon juice. Mash them lightly and set aside. In the bowl

Serves 8 | 7g protein

For the cakes:
1 cup white whole-wheat flour, plus more for dusting
1 cup spelt flour
1 tablespoon baking powder
2 tablespoons coconut sugar
½ teaspoon kosher salt
8 tablespoons unsalted butter, chilled, cut into ½-inch pieces
½ cup plus 2 tablespoons whole milk
1 large egg
1 tablespoon unsalted butter, melted, for brushing

For the topping:
1 quart strawberries
2 tablespoons coconut sugar
1 teaspoon fresh lemon juice
1 pint heavy cream
2 tablespoons confectioners' sugar
2 teaspoons pure vanilla extract

CONTINUED

of an electric mixer fitted with the whisk attachment, whip the cream until thickened but still soft. Stir in the confectioners' sugar and vanilla with a spoon.

To assemble the shortcakes, cut them in half horizontally with a knife. Spoon a good scoop of strawberries with their juice on the bottom of each biscuit. Top with the other half of the biscuit and spoon on more berries. Serve with a generous spoonful of the whipped cream.

Coconut Cream

A vegan alternative to whipped cream, this topping made from solid coconut cream is so decadent it's a dessert in itself. Because coconut milk has a good long shelf life, it's easy to keep a can in the fridge for when you might need to whip some up (as opposed to heavy cream, which spoils quickly).

Do not shake or stir the can of coconut milk before you open it because you want the cream to stay separated from the milk. After you do, scoop out the cream that is separated from the liquid and put it in the bowl of a stand mixer. Do not add any liquid if possible, just the solids.

Whip the solid cream on high speed until it's light and fluffy, 3 to 5 minutes. Stir in a drizzle of maple syrup and/or a teaspoon of vanilla, if desired.

Makes 1 cup

1 (25-ounce) can full-fat coconut milk, chilled in the refrigerator for at least 8 hours
Pure maple syrup and/or pure vanilla extract (optional)

Cook's Note:

Coconut milk that contains guar gum will not work. Guar gum keeps the milk and fat emulsified.

Brown Rice Berry Crisp

Thick and bubbly when warm from the oven, the filling of this crisp becomes almost pielike from the addition of brown rice. The oat flour makes the topping slightly crisp, but not crunchy like some traditional crisps. Yogurt or ice cream is de rigueur here. Leftovers, if any, make a good breakfast with a cup of steaming hot coffee.

Preheat the oven to 350°F. Grease a 9 x 13-inch baking dish and set aside.

In a large bowl, toss the berries and rice with the coconut sugar, oat flour, lemon zest and juice, and salt.

Make the topping: In a small bowl, stir together the topping ingredients until crumbly.

Spread the berry mixture evenly in the prepared pan and sprinkle with the topping. Bake for 25 minutes, or until the topping is bubbly and just beginning to turn golden. Serve warm with vanilla yogurt, ice cream, or Coconut Cream (page 201).

Serves 8 | 5 g protein

3 pints (6 cups) any combination of blueberries, blackberries, and raspberries
1 cup cooked short-grain brown rice
½ cup coconut sugar
1 tablespoon oat flour
1 teaspoon lemon zest
1 teaspoon fresh lemon juice
Pinch of kosher salt

For the topping:
¾ cup oat flour
½ cup old-fashioned rolled oats
½ teaspoon baking powder
½ teaspoon kosher salt
¼ cup coconut sugar
2 tablespoons pure maple syrup
¼ cup coconut oil or unsalted butter, melted
½ teaspoon ground cinnamon

Cook's Note:

If you need for this to be totally gluten-free, be sure your oat flour and rolled oats are both certified gluten-free, as conventional oat products may contain traces of gluten.

Matcha and Mango Chia Puddings

There are very few desserts that are this simple. The ingredient list is brief, but what you do need ample amounts of is time. It takes several hours in the refrigerator for the chia seeds to thicken the pudding and for the pudding to chill. Although not a last-minute dessert, it's an easy one.

Serves 4 | 4 g protein

For the matcha chia puddings:
2 cups coconut milk
2 teaspoons pure maple syrup
1 teaspoon matcha powder
⅓ cup chia seeds

For the mango chia puddings:
2 cups coconut milk (or your preferred milk, such as soy or almond)
2 teaspoons pure maple syrup
½ cup mango purée
⅓ cup chia seeds

Cook's Note:

Here I've given the option for matcha or mango flavorings, but you can also substitute any fruit purée, such as raspberry or strawberry, for the mango. You may also substitute another type of milk, such as almond or soy, for the coconut milk.

Put the coconut milk in a medium-size bowl. Stir in the maple syrup and either the matcha or mango purée. Add the chia seeds and stir again.

Divide the pudding between 4 (4-ounce or bigger) cups or dessert dishes. Chill the pudding for 6 hours or overnight. Serve chilled.

Peanut Butter Power Bites

No wonder peanut butter is America's favorite sandwich spread. Peanuts have more protein, niacin, and folate than any other nut (although actually a legume, peanuts are grouped with nuts in the US). Peanut butter is an inexpensive quality protein and thus a budget-friendly option for lunchboxes. It's also versatile. Here peanut butter is the base for these tiny, teaspoon-size bites that give a good power boost. These sweet little rounds are almost like a truffle. You can roll them in unsweetened cocoa powder to make them more trufflelike, or just eat them without the extra dusting.

Makes 24 bites | 15 g protein in 6 bites

6 dates (about ¼ cup)

¾ cup crunchy peanut butter

2 teaspoons honey

2 tablespoons mini semisweet chocolate chips

2 tablespoons vanilla soy milk

½ teaspoon pure vanilla extract

¼ cup old-fashioned rolled oats

1 to 2 tablespoons unsweetened cocoa powder (optional)

Cook's Note:

Use this recipe as a base for any nut butter, including sun butter if you are concerned with nut allergies. Also, look for gluten-free oats and chocolate chips if gluten is an issue.

Put the dates in a food processor and pulse several times until they are finely chopped. Add the peanut butter, honey, chocolate chips, soy milk, vanilla, and oats and run the processor until the dough comes together. Using a spatula, scrape down the sides of the food processor and stir the dough together.

Using a teaspoon, make small trufflelike balls. If desired, roll in cocoa powder. Store in an airtight container.

Pineapple Pops

Light and refreshing, these icy pops are a naturally sweet treat. Although they contain milk, they're more citrusy and fruity than creamy. Just a few licks and you may feel transported to somewhere a little tropical—that is, unless you're already there.

Put the pineapple and soy milk in a blender and process on high speed until smooth. Add the mint and the lime zest and juice and process for another few seconds.

Pour the mixture into six Popsicle molds and freeze until solid, about 4 hours.

Makes 6 (3-ounce) pops | 1 g protein

2 cups pineapple chunks (about
¾ pound)
¾ cup vanilla soy milk
¾ teaspoon chopped fresh mint
2 teaspoons lime zest
2 teaspoons fresh lime juice

Cook's Note:

If you don't have Popsicle molds, use an ice cube tray. Fill the tray with the blended pineapple mixture. Cover the tray with two or three layers of plastic wrap. Make a very small slit for each section and insert a Popsicle stick, pulling the wrap taut so the sticks stand upright and freeze.

FOOD SENSITIVITIES BY RECIPE

Look for the recipes in these categories if you have a food sensitivity or cook for someone who does

H, when honey is replaced with another sweetener

N-D, when non-dairy milk is used

G-F, when gluten-free bread or tortillas are used

N-C, when cheese (optional) is omitted

G-F G, when a gluten-free grain is used

E-L M, when an eggless mayonnaise is substituted for traditional mayonnaise

G-F B, when gluten-free beer is used

OO, when olive oil is substituted for the butter

GLUTEN-FREE

NOTE: When preparing a dish gluten free, always use gluten-free oats
to eliminate any cross-contamination with gluten

Sweet Potato Almond Milk Smoothie

Green Tea Pea Smoothie

Warming Breakfast Broth

Maple Granola Clusters

Brown Rice Chia Seed Porridge

Overnight Steel-Cut Oats

Creamy Amaranth Banana Porridge

Black Bean Breakfast Burrito, G-F

Root Vegetable Hash with Fried Eggs

Classic Egg and Cheese Sandwich, G-F

Kale and Carrot Tofu Scramble

Lentil, Spinach, and Tomato Frittata

Tarragon Egg Salad

Kale, Roasted Beet, and Edamame Salad

Farro, Baby Beet, and Pea Shoot Salad, G-F G

Crunchy Mung Bean Sprout Salad

Vegetable Noodles with Hemp Seed Basil Pesto

Warm Buckwheat Salad with Apples

TLT (Tempeh, Lettuce, and Tomato)
Sandwich, G-F

Brown Rice Avocado Collard Wraps

Roasted Eggplant and Zucchini Wraps, G-F

Grilled Vegetable and Fresh Ricotta
Sandwich, G-F

Favorite Tofu Bánh Mì, G-F

Vegan Zucchini Roll-ups

Mushroom and Truffle Oil Frittata

Creamy Cashew Soba Noodles

Brown Rice Balls (*Onigiri*)

Spicy Three-Bean Chili, G-F B

Stuffed Poblano Peppers

Roasted Kale, Green Bean, and Lentil
Bowl, G-F G

Chickpea Ratatouille in Parchment

Forbidden Black Rice with Pumpkin

Crispy Tofu Cauliflower Coconut Curry

Roasted Acorn Squash with Quinoa

Sea Vegetable Brown Rice Bowl

Red Kidney Bean Stew

Saag Paneer

Summer Tomatoes with Millet and Pesto Cream

Roasted Brussels Sprouts with Maple-Sugared
Orange Rind

White Bean Mash with Crispy Sage and
Brown Butter

Crispy Three-Grain Cake with Mozzarella
and Tomatoes

Portobello Mushrooms with Freekeh
and Artichokes, G-F G

Rich Lentil Stew

Jamaican Rice and Peas

Jamaican-Style Collard Greens

Red Lentil Hummus

Tamari Toasted Seeds

Roasted Paprika Chickpeas

Spinach and Chickpea Spoon Fritters

Roasted Edamame and Kale

Carrot Mash Tartine, G-F

Nut + Seed Protein Bars

Sriracha Deviled Eggs

Maple Baked Apples

Dark Chocolate Beet Cakes

Brown Rice Berry Crisp

Peanut Butter Power Bites

Matcha and Mango Chia Puddings

Pineapple Pops

Coconut Cream

VEGAN
NO EGGS NO DAIRY NO HONEY

Sweet Potato Almond Milk Smoothie

Green Tea Pea Smoothie, H

Warming Breakfast Broth

Brown Rice Chia Seed Porridge, N-D

Overnight Steel-Cut Oats, N-D

Creamy Amaranth Banana Porridge

Breakfast Barley with Raspberries and
Honey Figs, H

Kale and Carrot Tofu Scramble

No-Knead Dark Rye Bread

Kale, Roasted Beet, and Edamame Salad, H

Farro, Baby Beet, and Pea Shoot Salad, N-C, H

Crunchy Mung Bean Sprout Salad

Vegetable Noodles with Hemp Seed Basil Pesto

Creamy Mushroom Barley Soup, OO

TLT (Tempeh, Lettuce, and Tomato)
Sandwich, E-L M

Brown Rice Avocado Collard Wraps
Favorite Tofu Bánh Mì, E-L M
Vegan Zucchini Roll-ups
Creamy Cashew Soba Noodles
Brown Rice Balls (*Onigiri*)
Spicy Three-Bean Chili
Stuffed Poblano Peppers, N-C
Roasted Kale, Green Bean, and Lentil Bowl, H
Chickpea Ratatouille in Parchment
Crispy Tofu Cauliflower Coconut Curry
Roasted Acorn Squash with Quinoa, OO
Sea Vegetable Brown Rice Bowl
Red Kidney Bean Stew
Roasted Brussels Sprouts with Maple-Sugared
    Orange Rind, OO
Jamaican Rice and Peas
Jamaican-Style Collard Greens
Red Lentil Hummus
Seeded Crackers
Tamari Toasted Seeds
Roasted Paprika Chickpeas
Spinach and Chickpea Spoon Fritters
Roasted Edamame and Kale
Nut + Seed Protein Bars, H
Brown Rice Berry Crisp
Peanut Butter Power Bites, H
Matcha and Mango Chia Puddings
Pineapple Pops
Coconut Cream

DAIRY-FREE

Sweet Potato Almond Milk Smoothie
Green Tea Pea Smoothie
Warming Breakfast Broth
Maple Granola Clusters
Brown Rice Chia Seed Porridge, N-D

Overnight Steel-Cut Oats, N-D
Creamy Amaranth Banana Porridge
Breakfast Barley with Raspberries and
    Honey Figs, N-D
Quinoa Oat Breakfast Cookies
Root Vegetable Hash with Fried Eggs
Kale and Carrot Tofu Scramble
Tarragon Egg Salad
No-Knead Dark Rye Bread
Kale, Roasted Beet, and Edamame Salad
Farro, Baby Beet, and Pea Shoot Salad, N-C
Crunchy Mung Bean Sprout Salad
Vegetable Noodles with Hemp Seed Basil Pesto
Creamy Mushroom Barley Soup, OO
Brown Rice Avocado Collard Wraps
Favorite Tofu Bánh Mì
Vegan Zucchini Roll-ups
Creamy Cashew Soba Noodles
Brown Rice Balls (*Onigiri*)
Spicy Three-Bean Chili
Stuffed Poblano Peppers, N-C
Roasted Kale, Green Bean, and Lentil Bowl
Chickpea Ratatouille in Parchment
Crispy Tofu Cauliflower Coconut Curry
Roasted Acorn Squash with Quinoa, OO
Sea Vegetable Brown Rice Bowl
Red Kidney Bean Stew
Jamaican Rice and Peas
Jamaican-Style Collard Greens
Red Lentil Hummus
Seeded Crackers
Tamari Toasted Seeds
Roasted Paprika Chickpeas
Spinach and Chickpea Spoon Fritters
Roasted Edamame and Kale
Nut + Seed Protein Bars
Scallion Sesame Pancakes

Spiced Squash Snack Cake

Brown Rice Berry Crisp

Peanut Butter Power Bites

Matcha and Mango Chia Puddings

Pineapple Pops

Coconut Cream

EGG-FREE

Sweet Potato Almond Milk Smoothie

Green Tea Pea Smoothie

Warming Breakfast Broth

Brown Rice Chia Seed Porridge

Overnight Steel-Cut Oats

Creamy Amaranth Banana Porridge

Breakfast Barley with Raspberries and
    Honey Figs

Kale and Carrot Tofu Scramble

No-Knead Dark Rye Bread

Kale, Roasted Beet, and Edamame Salad

Farro, Baby Beet, and Pea Shoot Salad

Crunchy Mung Bean Sprout Salad

Vegetable Noodles with Hemp Seed Basil Pesto

Warm Buckwheat Salad with Apples

Creamy Mushroom Barley Soup

Brown Rice Avocado Collard Wraps

Grilled Vegetable and Fresh Ricotta Sandwich

Favorite Tofu Bánh Mì, E-L M

Vegan Zucchini Roll-ups

Creamy Cashew Soba Noodles

Brown Rice Balls (*Onigiri*)

Spicy Three-Bean Chili

Stuffed Poblano Peppers

Roasted Kale, Green Bean, and Lentil Bowl

Chickpea Ratatouille in Parchment

Crispy Tofu Cauliflower Coconut Curry

Roasted Acorn Squash with Quinoa

Sea Vegetable Brown Rice Bowl

Red Kidney Bean Stew

Summer Tomatoes with Millet and Pesto Cream

Roasted Brussels Sprouts with Maple-Sugared
    Orange Rind

White Bean Mash with Crispy Sage and
    Brown Butter

Crispy Three-Grain Cake with Mozzarella
    and Tomatoes

Portobello Mushrooms with Freekeh
    and Artichokes

Rich Lentil Stew

Jamaican Rice and Peas

Jamaican-Style Collard Greens

Red Lentil Hummus

Seeded Crackers

Forbidden Black Rice with Pumpkin Wedges

Tamari Toasted Seeds

Roasted Paprika Chickpeas

Spinach and Chickpea Spoon Fritters

Roasted Edamame and Kale

Carrot Mash Tartine

Nut + Seed Protein Bars

Maple Baked Apples

Brown Rice Berry Crisp

Peanut Butter Power Bites

Matcha and Mango Chia Puddings

Pineapple Pops

Coconut Cream

Acknowledgments

This book is representative of community and its inherent goodness. Akin to a barn raising, so many pitched in—quite a few from our Catskill valley—to make the pages of this book happen. I thank every one of you who contributed in any way, big or small.

Joshua Holz, I am in awe of your dedication, keen eye, and crazy-fast photo-editing skills. My deepest thanks for bringing beauty to every image. Jen, you are a one of a kind girl! No words can convey my gratitude for your support. George, I will forever cherish your punny self, and am so grateful you got us to chase the light. I am indebted to you, Team Holz.

Jeri Heiden, you are the most pleasant person to work with—ever. Thanks for your unwavering support; your designs are exquisite. Rebecca Shim, you masterfully swooped in with your champion kitchen skills and made everything right. Thank you for generously sharing your knowledge and recipes. Griselda Meija, you are the best! Thanks for your tireless work and assistance—and always being there, which can be said of you too, 3D. From the time you could schlep a prop, you've been the best on- (and off-) set assistant (and companion); you find a laugh in everything. I'd also like to thank NYUers: Becky Blair Hughes for your contributions and input; (or should I say Tisch'er) Justin Lanier, I'm sorry you were away but appreciate the few images you contributed.

A thousand thanks to my master recipe tester (and coordinator) Michelle Perreault-Dougherty. I adore you for so many reasons, including your determination and good humor. A huge thanks to my other diligent testers as well: Claudia Haj Ali, Dani Rasmus Crichton, Charlotte Haukedal, Robin Kornstein, Amy Jackson, Joan Tsianos, and Anni Gundeck.

I'm also deeply grateful to my dedicated nutritional analysis team: Lily Davis Bosch, Camilla Ffrench, and Anni Gundeck. Your keen insights are duly noted.

There are a handful of people I'd like to thank for coming down to Holz Farm: Robin and Jasmine Chess; Chris, Ben, Dylan, and Brendan Dougherty; Michael Graytser and Madeleine Hale, Jazarah and Amanuel Shim. And to other WVers: Julia Rose for your turnips, and MJ and Rolf Reiss, for always, *always* opening your doors and being the incredible people you are.

I'd like to thank Migliorelli Farm and the crew there for continually providing fresh and delicious food-stuffs. I also want to give a shout out to Sean at my favorite kitchen store bluecashew in Rhinebeck, NY, for graciously loaning me some of your beautiful dishes.

A special thanks to the extremely generous Carrie Bachman for your support and ideas. I am also grateful to Sarah Copeland for loaning me your gorgeous dishes and surfaces, and for your guidance and wisdom, answering random questions as they arose.

And to everyone at Countryman Press—I am extremely appreciative of your efforts: my fantastic editor Ann Treistman, for your patience and insight; Sarah Bennett, for the multitude of things you do; Devon Zahn and Natalie Eilbert for implementing change after change; the sales team for getting this book out there, and Devorah Backman for your brainstorming sessions and drive to make things happen.

I am grateful to Kayleen St. John, MS RD, for your nutritional counsel and also to you, and the Natural Gourmet Institute, for your health-supportive culinary training. Thanks, too, to my lab partner Ankita Kastia Sancheti, for teaching me about Indian cooking.

I am indebted to Kyle Gerry, my DL, for the many hours you devoted to reading the various versions of the book. No price can be put on your eagle eye and thoughtful feedback.

To the fabulous Sharon Bowers, my crackerjack agent, I cherish you beyond belief.

To my beloved sister, Lisa, for always taking the time to pitch in—you're the best!

To my Mom and Dad for always being there and showing me the value of family.

To Anna and Camilla—you are my joy. I treasure your intelligence and input, and willingness to always help out. Plus your humor and good appetites—I love girls who know how to eat well! Finally, to Jim, my heart is yours. I am constantly amazed by your endless encouragement and support—and willingness to run to the store. I value all you teach me—and your very witty quips that keep us smiling every day.

Index